ACS SYMPOSIUM SERIES **304**

Fungicide Chemistry

Advances and Practical Applications

Maurice B. Green, EDITOR
Independent Consultant

Douglas A. Spilker, EDITOR
Mobay Corporation

Developed from a symposium sponsored by
the Division of Pesticide Chemistry
at the 188th Meeting
of the American Chemical Society,
Philadelphia, Pennsylvania,
August 26–31, 1984

American Chemical Society, Washington, DC 1986

Library of Congress Cataloging-in-Publication Data
Fungicide chemistry.
  (ACS symposium series; ISSN 0097-6156; 304)
  Includes indexes.
  1. Fungicides—Congresses. 2. Fungi,
Phytopathogenic—Control—Congresses.
  I. Green, Maurice B. (Maurice Berkeley) II. Spilker,
Douglas A., 1953–   . III. American Chemical
Society. Division of Pesticide Chemistry. IV. American
Chemical Society. Meeting (188th: 1984: Philadelphia,
Pa.) V. Series.

SB951.3.F84   1986   632'.952   86-1248
ISBN 0-8412-0963-4

ACS Symposium Series

M. Joan Comstock, *Series Editor*

Advisory Board

FOREWORD

The ACS SYMPOSIUM SERIES was founded in 1974 to provide a medium for publishing symposia quickly in book form. The format of the Series parallels that of the continuing ADVANCES IN CHEMISTRY SERIES except that, in order to save time, the papers are not typeset but are reproduced as they are submitted by the authors in camera-ready form. Papers are reviewed under the supervision of the Editors with the assistance of the Series Advisory Board and are selected to maintain the integrity of the symposia; however, verbatim reproductions of previously published papers are not accepted. Both reviews and reports of research are acceptable, because symposia may embrace both types of presentation.

CONTENTS

PREFACE

Pesticides, unlike many advances in technology that are based on original work conducted by universities, originated in the chemical industry. New pesticides with commercial applications are still discovered almost entirely by that industry.

The recipient of the Burdick and Jackson Award for Excellence in Pesticide Chemical Research is selected each year by an international committee of the Pesticide Chemistry Division of the American Chemical Society. This prestigious award is given to recognize sustained outstanding contributions to the science of pesticide chemistry. Since its inception in 1969, it has been presented to a succession of eminent scientists from universities and government research stations. In 1984 the winner was, for the first time, a man who had spent his entire working life in the agrochemicals industry—Karl Heinz Büchel of Bayer AG of the Federal Republic of Germany.

The award to Karl Heinz Büchel honors agricultural chemical research in industry throughout the world and particularly the German agrochemical industry that has made so many contributions starting with the time of Gerhard Schrader.

This volume contains the presentations given by eminent members of the scientific community to honor Karl Büchel. Experts from the United States and Europe were brought together to discuss aspects of the chemistry of azole fungicides and ways in which this chemistry is being used for disease control in major food crops.

MAURICE B. GREEN
1 Dinorben
79/81 Woodcote Road
Wallington
Surrey SM6 0PZ
England

DOUGLAS A. SPILKER
Mobay Corporation
Agricultural Chemicals Division
8400 Hawthorn Road
Kansas City, MO 64120

November 13, 1985

The History of Azole Chemistry

Karl H. Büchel

Bayer AG, Forschung und Entwicklung, D5090 Leverkusen, Federal Republic of Germany

Since the earliest days of agriculture, insect pests, weeds and plant diseases have been some of the major problems of agriculture. Insect pests are visible and could, at least in some cases be countered by hand removal. A certain level of weed elimination was achieved by hoeing and hand weeding, an never ending task. However, rust, powdery mildew and smut being invisible enemies, spread throughout the fields like an unpredictable fate. Thus they occupied the imagination of rural folk, and magical concepts of disease control predominated in the early days of agriculture.

Until the discovery of Bordeaux mixture in 1880, farmers had no real possibility of defending their crops against the ravages of fungal diseases. Fungi had only been identified a few years previously as the cause of plant diseases. Apart from a few empirical measures to prevent disease, active control was not possible. Massive disease epidemics often had catastrophic social consequences. The legacy of the Irish potato famine of the 1840's can be seen even today. It is difficult to imagine the significance of the pioneer fungicides, ones based on copper, sulphur and mercury, had in their time.

The organic fungicides of the dithiocarbamate and phthalimide type (e.g. Captan) were a breakthrough in this field in the nineteen thirties and forties. Although they only have protective activity and thus must be used prophylactically, they found broad applications due to their high plant compatibility and broad disease control spectrum.

A further milestone in the development of fungicides was the discovery of the so-called systemic fungicides, chemicals that are taken up by the plant and transported within it. The fungicide classes found in the sixties, including the oxathiines, pyrimidines and organophosphates, are characterized by being absorbed by the leaves, often also by the seeds and roots, and being transported acropetally within the plant. These products have only a very narrow disease control spectrum. The oxathiines are active against Basidiomycetes, mainly against rusts and smuts; the pyrimidine derivatives are active against powdery mildews. The organophosphates are used for Pyricularia control in rice.

0097–6156/86/0304–0001$07.00/0
© 1986 American Chemical Society

The much wider disease control spectrum of the benzimidazole fungicides (eg. benomyl, BCM, thiabendazole) - permitted far wider usage. In the beginning, these were suitable for control of numerous plant diseases, but a new phenomenon soon emerged - resistance! Due to the specific mode of action of these fungicides, resistance could appear quite rapidly. The conventional fungicides previously used had a broad biocidal activity and resistance had never been experienced.

History of Azole Chemistry

With the class of 1-substituted imidazoles and 1,2,4-triazoles, we found a new group of highly active fungicides and antimycotics (1,3). Since their discovery in the late sixties, several compounds from this chemical class have been commercially developed and successfully used for the control of plant diseases and for the treatment of human fungal infections. These so-called "azole fungicides and antimycotics" have set new standards in medicine and agriculture with respect to efficacy and range of disease control spectrum. Among this group, we find the most active compounds known today for control of plant diseases and human mycoses.

"Carbocation Hypothesis". Our work on N-substituted imidazoles and triazoles began in the mid-sixties with a very simple hypothesis. It was known that tropylium compounds, cycloheptatriene derivatives and some N-tritylamines had biological activity. For example N-tritylmorpholine (Frescon; Shell) had been developed as a molluscicide and used for control of water snails (Figure 1). The activity of some triphenylmethane dyestuffs against certain endoparasites and the biological activity of naturally occurring tropolone derivatives had also been reported (3).
From the viewpoint of a chemist, these systems all have one feature in common: they are all able to form relatively stable carbonium ions. To explain the biological activity of these compounds, we speculated at that time that their corresponding carbonium ions could possibly interfere in the metabolic processes of biological systems, for example in protein metabolism. The N-tritylimidazoles and triazoles seemed suitable to us since they have a tendency to form stable carbocations (Figure 2).
Analogous to the N-acylimidazoles, in which the enhanced reactivity of the acyl group towards nucleophiles is favoured by inclusion of the amide-nitrogen lone electron pair in the $\triangle$-system of the heterocycle, nucleophilic attack on the trityl part of the N-tritylazoles also should be a facile process (11).

N-Tritylazoles, Diphenylmethylazoles and Related Compounds. On the basis of this hypothesis, we began the synthesis of N-tritylimidazoles and sent them for biological testing. Very shortly, the excellent activity of this type of compound against human and plant pathogenic fungi and yeasts was evident in our biological screens. The activity against powdery mildew fungi was particularly remarkable.
Following these initial, highly encouraging results, we cooperated with the biologists and mycologists, to try to optimize the

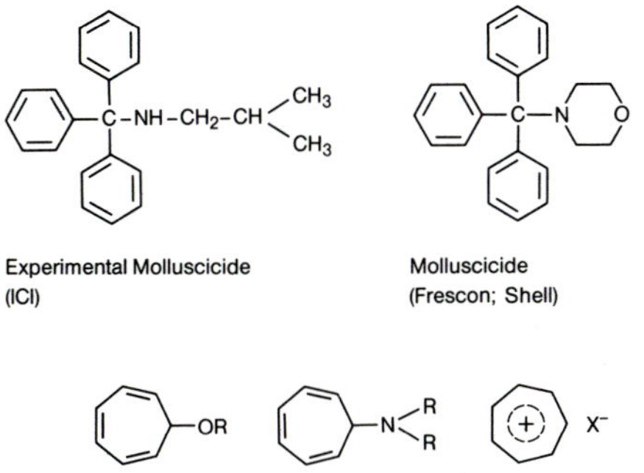
C-NH-CH2-CH
CH3
CH3
C
N
O
Experimental Molluscicide
(ICI)
Molluscicide
(Frescon; Shell)
OR
N
R
R
+
X⁻

Figure 1. Biologically active trityl-, cycloheptatrienyl- and tropylium derivatives.

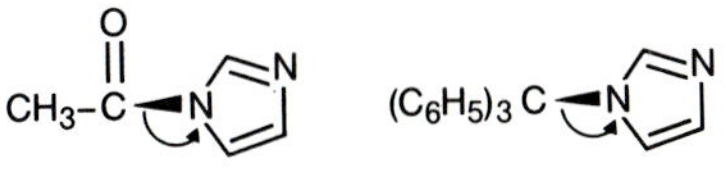
O
CH3-C
N
N
(C6H5)3 C
N
N

Figure 2. N-Acetyl- and N-trityl-imidazole.

antimycotic and fungicidal activity of these tritylazoles. This was
accomplished through a wide variation of the substitution pattern in
the tritylazoles, and then the identification of the relationship
between structure and activity (3). Many structural variations were
made on the basic 1-tritylimidazole skeleton (Figure 3). Of the
azole components studied, imidazole and 1,2,4-triazole were found to
be the structural elements essential for activity. The corresponding
N-substituted derivatives of pyrazole, tetrazole, benzimidazole, as
well as C-substituted imidazoles and triazoles, had much weaker or
lacked fungicidal activity. In the case of 1,2,4-triazole, only
substitution at position 1 resulted in compounds with good antifungal
properties; the corresponding isomeric 4-substituted triazole derivatives were much less active.

In contrast, the trityl residue can be widely varied without
loss of biological activity. Neither varying the substitution
pattern in the aryl residues of the trityl group nor replacing one or
two of the aryl rings by other substituents, leads to loss of the
basic antifungal properties. Substitution of the aromatic rings by
alkyl, cycloalkyl, alkenyl and alkinyl substituents, or the introduction of ester groups, acyl residues and heterocycles into the original trityl system also leads to compounds with excellent activity.
The system can be varied so widely that one can even replace one of
the phenyl rings in the trityl group with hydrogen without loss of
fungicidal activity.

From the tritylazole group, clotrimazole (Canesten, Mycelex,
Lotrimin, Empecid) was developed and successfully introduced into
medical practice as a broad spectrum antimycotic for the topical
treatment of fungal infections. In the same class the 1,2,4-triazole
derivatives were found to be superior to the corresponding imidazoles
for the control of phytopathogenic fungi. Fluotrimazole (Persulon),
a non-systemic fungicide, was developed for the control of powdery
mildew on cereals and fruit (Figure 4).

Two products originate from the diphenylmethylimidazole class.
Apart from its fungicidal properties, lombazole also has good activity against Propionobacterium acne and is used as the active ingredient in anti-acne preparations. Bifonazole (Mycospor), the halogen-
free analogue of lombazole, is a topical broad-spectrum antimycotic
with a long retention period in the dermis. Bifonazole has shown
excellent control of tinea corporis and tinea pedis of humans caused
by Trichophyton species (Figures 5-7) (3,6,10).

At a very early stage, we carried out investigations on the
relationship between antimycotic activity and chemical structure of
the N-tritylimidazoles. In particular, we hoped to identify a
positive correlation between rate of hydrolysis and biological
activity, and therefore obtain support for our original working
hypothesis. However, we found very quickly that there was no relationship between the rate constants for acid hydrolysis, as a measure
of the tendency to form a trityl cation, and the corresponding
antimycotic activity. This means that our concept had led to the
desired result - the discovery of a new class of biologically active
compounds based on a highly speculative hypothesis. Today, after
more than 15 years of research work and intensive investigations on
their mode of action, we know that the azole fungicides belong to the
large group of ergosterol biosynthesis inhibitors and interfere with
the biosynthesis of fungal steroids.

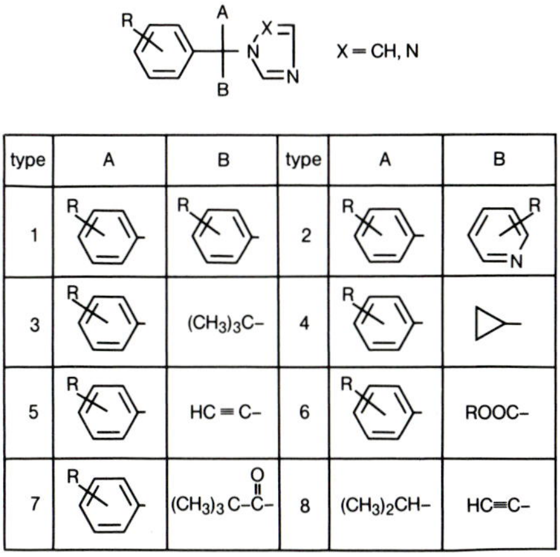

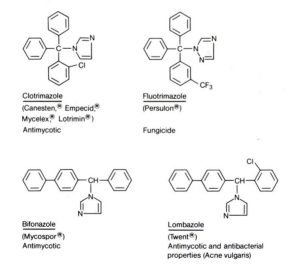

Figure 3. Variability in the class of N-tritylazoles. (Reproduced with permission from Ref. 9. Copyright 1983 Pergamon Press.)

Clotrimazole
(Canesten,® Empecid,®
Mycelex,® Lotrimin®)
Antimycotic

Fluotrimazole
(Persulon®)

Fungicide

Bifonazole
(Mycospor®)
Antimycotic

Lombazole
(Twent®)
Antimycotic and antibacterial
properties (Acne vulgaris)

Figure 4. N-trityl- and N-diphenyl-methyl-azoles developed for practical use.

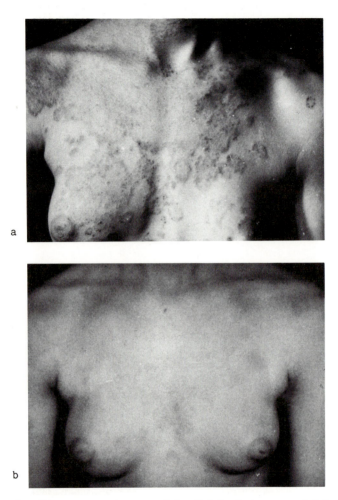

Figure 5. Tinea corporis caused by <u>Trichophyton</u> sp. (a) before and (b) after topical treatment with <u>bifonazole</u> (one treatment daily for three weeks).

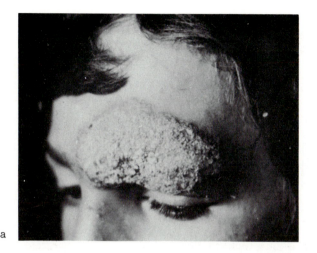

a

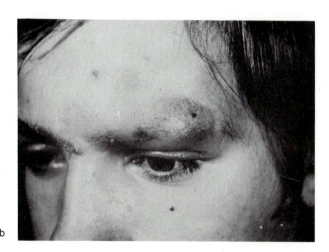

b

Figure 6. Tinea corporis caused by <u>Trychophyton</u> sp. (a) before and (b) three weeks after topical treatment with bifonazole.

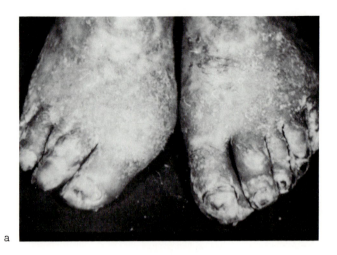

a

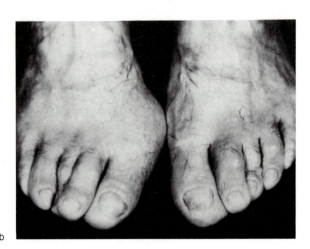

b

Figure 7. Tinea pedis (infection of feet) caused by Trycho-
phyton sp. (a) before and (b) after therapy with bifonazole.

Syntheses in the Trityl Group. During our synthetic work in the trityl group we developed a series of alternative synthetic routes, each with its own particular utility depending on the substitution pattern of the trityl component. The classical method of N-alkylation of azoles with a trityl chloride or tritylium salt as well as the use of activated azoles, such as trimethylsilylazole or an acylimidazole was often found to be the optimum method for creating the C-N bond under moderate conditions (Figure 8). In the reaction of trityl carbinols or other tertiary alcohols with the azolides of inorganic acids, such as thionyl-, carbonyl-bis-imidazole or phosphoric acid tris-imidazolide, we found a new method for the preparation of 1-substituted imidazoles. This was particularly valuable for the synthesis of azole derivatives with certain functional groups where the other alkylation methods failed. Two of the most different synthetic routes are illustrated by the synthesis of fluotrimazole (Persulon) and of a N-propinylimidazole (BAY d 9603) (Figures 9-10) (9).

The synthesis of fluotrimazole starts from m-xylene. Peroxide catalyzed perchlorination converts this to m-trichloromethyl-benzotrichloride. m-Trichloromethyl-benzotrifluoride is then obtained by selective chlorine/fluorine exchange. This key product is also readily accessible on a technical scale by conproportionation of the two corresponding m-trihalomethyl-benzotrihalogenides. Friedel-Crafts reaction with benzene leads to trifluoromethyl-tritylchloride, which reacts smoothly with 1,2,4-triazole in polar solvents to give fluotrimazole.

N-propinylazoles of the BAY d 9603 type have a close structural relationship to the tritylazoles. They have a wide range of activity, particularly effective against mould fungi. Starting material here is the readily available diphenyl-propinol obtained by ethinylation of benzophenone. The method for the conversion of this carbinol into the corresponding N-substituted imidazole is the reaction with carbonyl or thionyl-bis-imidazole, or preferably with the tris-imidazolide of phosphoric acid in a polar solvent. All other alkylation methods in which carbocations occur as intermediates completely fail.

Phenethylimidazoles. During a search for hypnotic agents in the late sixties (2), a research group at Janssen Pharmaceutica discovered the antimycotic and fungicidal properties of another class of imidazole derivatives, the phenethylimidazoles. This class, which differs greatly from the tritylazoles with respect to the substituents at the azole nitrogen, now includes the antimycotic miconazole, and ·the fungicide imazalil. Imazalil was developed especially for control of Helminthosporium diseases of plants. Together with clotrimazole and fluotrimazole, miconazole and imazalil belong to the first generation of azole fungicides and antimycotics to reach the market (Figure 11).

Azolyl-O,N-acetals. Significant progress was achieved in the control of plant diseases with the discovery of the highly active class of so-called "triazolyl-O,N-acetal" fungicides at the beginning of the nineteen seventies. These compounds are O,N-acetals of 2-ketoaldehydes or 2-hydroxyaldehydes in which the triazole residue forms a part of the O,N-acetal function.

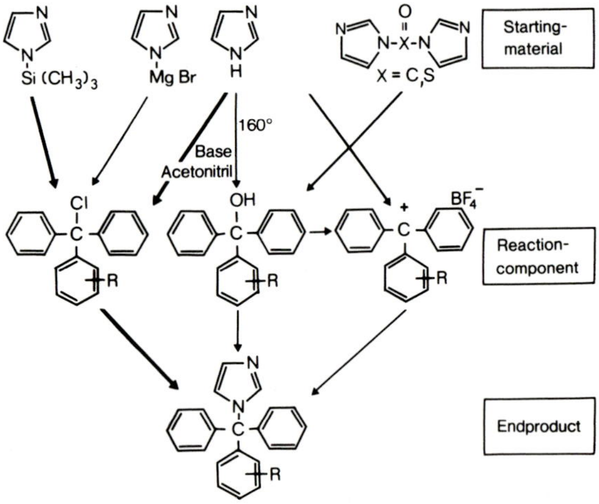

Figure 8. Routes to N-tritylimidazoles.

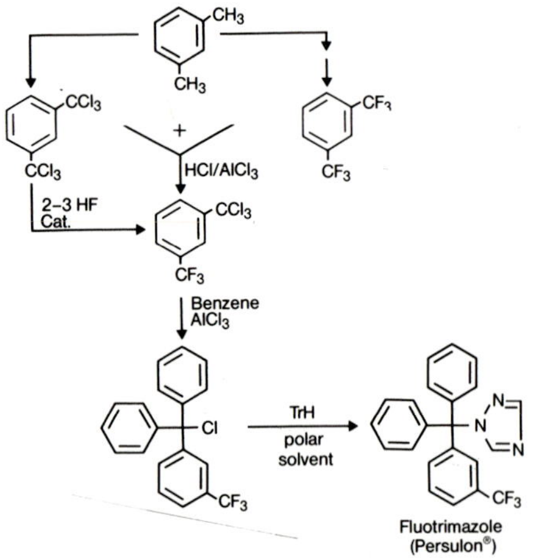

Figure 9. Pathways to fluotrimazole (Persulon). (Reproduced with permission from Ref. 9. Copyright 1983 Pergamon Press.)

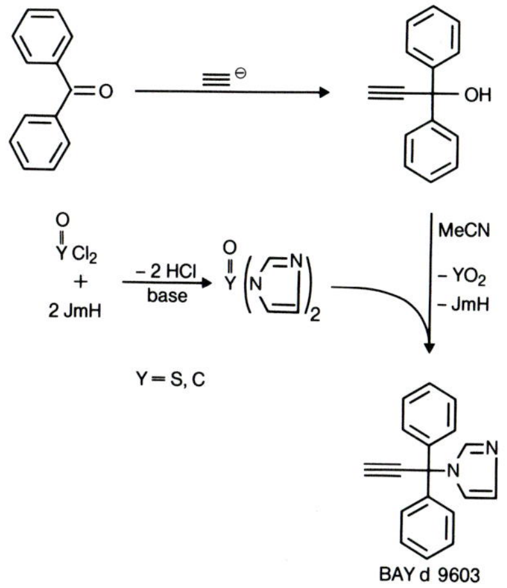

Figure 10. Synthesis of N̲-propinylazoles--an example. (Reproduced with permission from Ref. 9. Copyright 1983 Pergamon Press.)

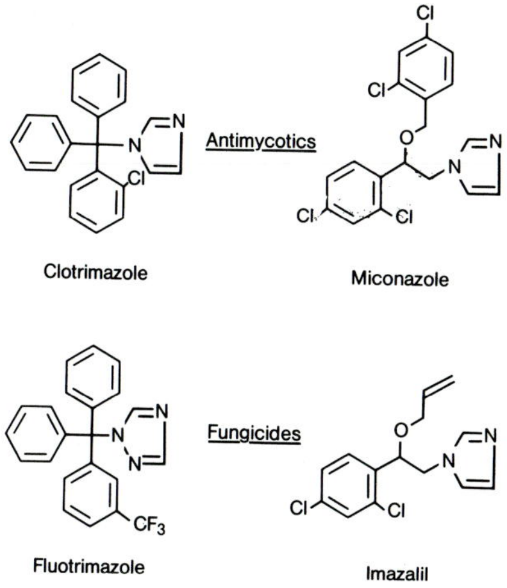

Figure 11. First azole antimycotics and fungicides developed for practical use. (Reproduced with permission from Ref. 9. Copyright 1983 Pergamon Press.)

The synthetic routes to the systemic fungicide triadimefon (Bayleton) and the cereal seed dressing triadimenol (Baytan), starting from pinacolone, are representative for this very interesting group of azole derivatives (Figure 12). For the first time, we were able to synthesize azole fungicides that were systemic in plants and transported mainly upward in the direction of growth. Triadimefon is a potent systemic fungicide with particularly high activity against powdery mildew and rust fungi. It is used in many different crops, but mainly in cereals and fruit. Triadimenol has excellent systemic properties making it suitable not only for the control of seed- and soil-borne fungal organisms but even of infections by wind-borne pathogens.

Other important members of the azolyl-O,N-acetal family that have been marketed thus far are bitertanol (Baycor, Sibutol) and the imidazole derivative climbazole (Baypival). Bitertanol is not systemic, but penetrates plant tissue and thus possesses curative and eradicative properties combined with protective activity. It is used for control of foliar diseases of various crops such as tree fruit, peanuts and bananas.

Climbazole is not an agricultural fungicide. It has a completely different antifungal profile than its triazole analogue, triadimefon, and has particularly high activity against mould fungi, yeasts and dermatophytes. It also has excellent activity against Pityrosporum ovale and is thus used as an active ingredient in anti-dandruff formulations (Ceox).

These examples (Figure 13) clearly show that, although there are very close chemical relationships within a class of compounds, there may be remarkable differences in the biological and biophysical properties. This enables the use of various compounds for specific purposes although their biological spectrums may overlap (4,5,7,8,9).

Chemical Variability of the Azole Class

The chemical variability of the 1-substituted azoles can already be recognized from the subgroups presented thus far, including the N-tritylazoles and their analogues, the N-diphenylmethylazoles, the phenethylazoles and the azolyl-O,N-acetals. The only essential feature is the imidazole or 1,2,4-triazole ring. The quality of the fungicidal action and of the properties necessary for practical use are determined only by the selection of a suitable substituent R.

A simple calculation can serve to demonstrate this chemical variability (Figure 14). If one designates the substituents of the N-substituted azoles as X, Y, Z and the number of variations as n, then the total number N of azole derivatives possible can easily be calculated. If n = 200, which is certainly a very modest number of different X, Y, Z substituents, then this means that with imidazole and triazole as components, it is possible to make 5.4 million different compounds!

One of the special features of azole compounds is the abundance of synthetic possibilities and therefore the abundant selection opportunities for biology and medicine. By examination of the properties of azole derivatives, one can recognize other applications for them apart from the control of fungi pathogenic to plants and man. Some of the azole derivatives, such as lombazole, exhibit

Cl_2; Br_2;
2 Cl_2
Cl (Br)
Cl
Cl
Cl
O⊖
TrH
1. Br_2 or SO_2Cl_2
2. TrH/K_2CO_3
Cl
Cl
OH
$NaBH_4$
Triadimefon
Cl
Triadimenol

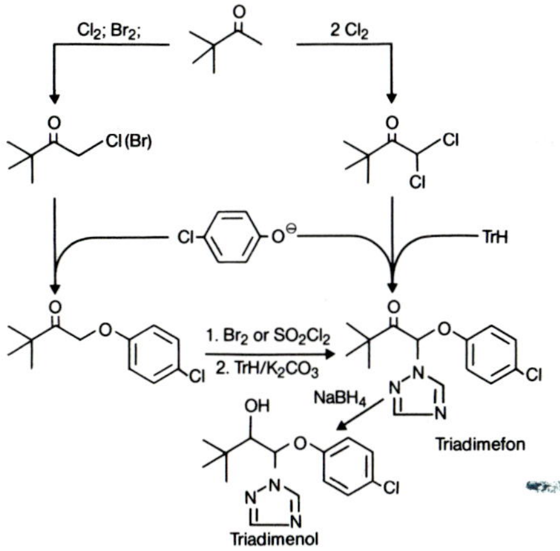

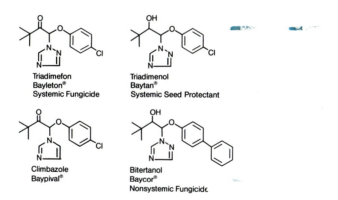

Figure 12. Pathways for the synthesis of triadimefon and triadimenol. (Reproduced with permission from Ref. 9. Copyright 1983 Pergamon Press.)

Triadimefon
Bayleton®
Systemic Fungicide
Triadimenol
Baytan®
Systemic Seed Protectant
Climbazole
Baypival®
Bitertanol
Baycor®
Nonsystemic Fungicide

Figure 13. Azolyl-O,N-acetals. (Reproduced with permission from Ref. 9. Copyright 1983 Pergamon Press.)

activity against gram-positive bacteria. Certain azole compounds interfere with the biosynthesis of gibberellins and influence the morphogenesis of green plants, indicating their possible use as plant growth regulators. It comes as no surprise that in the years following the recognition of the extraordinary biological properties of the azoles, intensive research work was initiated worldwide in many laboratories on imidazole and triazole derivatives.

Azole Compounds from other Classes

To date, this worldwide activity has led to a whole series of highly active azole derivatives developed for practical use in medicine and agriculture (Figure 15).
    Triazole derivatives of the dichlobutrazole and paclobutrazole types are closely related chemically to the triazolyl-O,N-acetals previously mentioned. The oxygen atom of the acetal function was merely replaced by a methylene group. Dichlobutrazole (ICI) was developed for the control of powdery mildew, rust and scab in cereals. Paclobutrazole is under development as a growth regulator.
    Propiconazole and etaconazole, two triazole derivatives discovered by Janssen and developed under license by Ciba-Geigy for control of fungal diseases in cereals and fruit, belong to another azole subgroup, the azolylmethyl-dioxolanes. Their chemical genesis from the phenethylazoles of the imazalil and miconazole type is easily recognizable. Ketoconazole (Janssen), an oral antimycotic introduced into human therapy, also belongs to this group.
    The imidazole derivative, prochloraz, has a peculiar feature compared to the azole derivatives. The nitrogen atom in the 1-position of the imidazole is bonded to a carbamoyl group and not to alkyl. This compound, which is regarded as a tetrasubstituted urea, has a very broad spectrum and can be used in cereals both as a seed dressing and a foliar fungicide (9).

Biological Activity and Stereochemistry

The N-vinyl-azole class, in which the 1-N atom of the azole is directly bonded to an sp^2-hybridized carbon of a substituted olefin, belongs to the relatively recently synthesized subgroup of bioactive azoles. Compounds of this nature can be obtained by condensation of aldehydes with triazolyl-pinacolone resulting in a mixture of the E and Z isomers of the corresponding α, β-unsaturated ketones. Borohydride reduction easily converts these ketones into the corresponding alcohols (9)(Figure 16).
    In these compounds, there is a marked relationship between molecular geometry and biological activity. From values reported in the literature and according to our own studies, the E isomers, in which the residue originating from the aldehyde is in the transposition to the triazole, are markedly superior to the Z isomers in their biological activity. By suitable control of the reaction conditions, it is possible to achieve an almost complete isomerization to the unsaturated E-triazolylketones. Subsequent reduction leads to the more active E-alcohols. This group of N-vinylazoles includes the triazole derivative S 3308 (Sumitomo), currently under development as

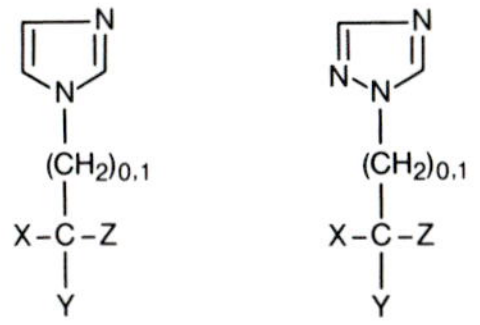

$$N=4\,\frac{n}{3} + \frac{n^2}{2} + \frac{n^3}{6}$$

N=Number of possible active compounds in the azole group
[for imidazoles and 1,2,4-triazoles]
n=Number of different substituents X, Y, and Z

n	10	50	100	200
N	880	88,400	688,800	5,413,600

Figure 14.   The chemical variability of the azole class.

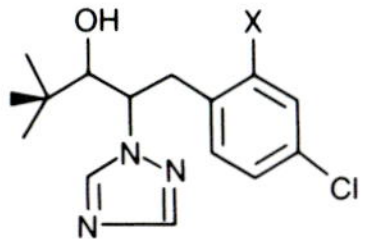

Diclobutrazole   (X = Cl)
(Fungicide; ICI)

Paclobutrazole   (X = H)
(Plant growth regulator; ICI)

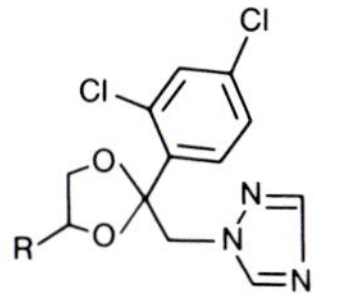

Etaconazole   (R = C_2H_5)
Propiconazole (R = C_3H_7)
(Fungicides; Ciba-Geigy/Janssen)

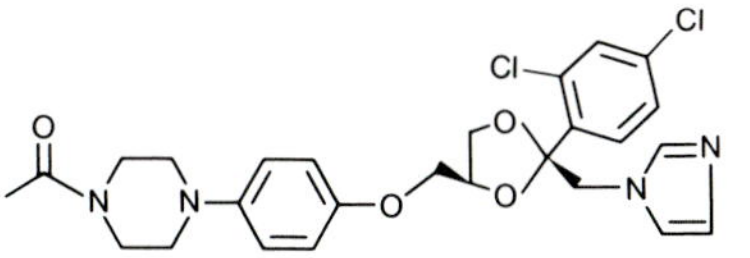

Ketoconazole
(Antimycotic; Janssen)

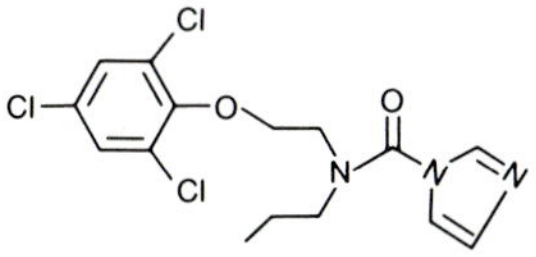

Prochloraz
(Fungicide; Boots)

Figure 15.   Azole fungicides and antimycotics from other subgroups.

a fungicide, as well as the growth regulators S 3307 and triapenthenol (RSW 0411, Bayer).

As in the use of dichlobutrazole and paclobutrazole, the biological spectrum of the two experimental compounds S 3307 and S 3308, whether fungicide or growth regulator, is dictated only by the substitution pattern in one part of the molecule. Triapenthenol, which is of particular interest as a growth regulator in rice and rape, also possesses a marked fungicidal activity.

Two enantiomers of triapenthenol were obtained from the racemate via diastereomeric ester derivatives (Figure 17). In the biological testing of the enantiomers, a clear separation of the biological properties was observed. The (-)-S-enantiomer is almost solely responsible for the growth regulator properties; the (+)-R-enantiomer has a markedly higher fungicidal activity (Figures 18-19).

Many of the azole fungicides mentioned previously exist as two geometrical isomers, that is two corresponding enantiomer pairs with 1R, 2R / 1S, 2S and 1R, 2S / 1S, 2R configurations, due to the presence of two chiral centres in the molecule. Examples of this are the triazole compounds etaconazole, propiconazole, dichlobutrazole, bitertanol and triadimenol. In the synthesis of triadimenol, the second asymmetric center is introduced by the reduction of the keto function in triadimefon. This results in the formation of two diastereomeric forms which have been found to have different fungicidal activities (Figure 20). The chemist is then faced with the task of developing stereoselective syntheses, in order to obtain a product with the highest possible proportion of the more active diastereomer. In the conversion of triadimefon to triadimenol, it was possible to steer the reaction in the desired direction by careful selection of the reduction system (9).

From studies of the biological activity of the individual enantiomers, we obtained a detailed insight into the relationship between absolute configuration of the triazolyl-O,N-acetals and their fungicidal properties. R- and S-triadimefon were obtained in high optical purity (99% e.e.) from racemic product by separation of its diastereomeric -bromocamphorsulphonate salts; the absolute configurations being determined by X-ray crystallography. Easily separable diastereomeric mixtures of enantiomeric triadimenols are then obtained by reduction of the two triadimefon enantiomers, the configuration at position 1 being preserved (Figure 21). Although R- and S-triadimefon exhibit practically no differences in activity within the accuracy limits of biological testing, the four enantiomeric triadimenols show marked gradations in their activity spectra. The highest fungicidal activity of all 4 enantiomers residues was with the (-)-1S, 2R enantiomer.

A similar relationship between absolute configuration and fungicidal activity was also observed with the enantiomers of bitertanol. As with triadimenol, the fungicidal activity of the (-)-1S, 2R enantiomer is markedly greater than that of the other 3 enantiomers (Figure 22).

Differences in the biological properties of the enantiomers have also been reported for other azole products that possess one or more chiral centers in the molecule. These few results clearly show that, as in many other biologically active classes of compounds, the

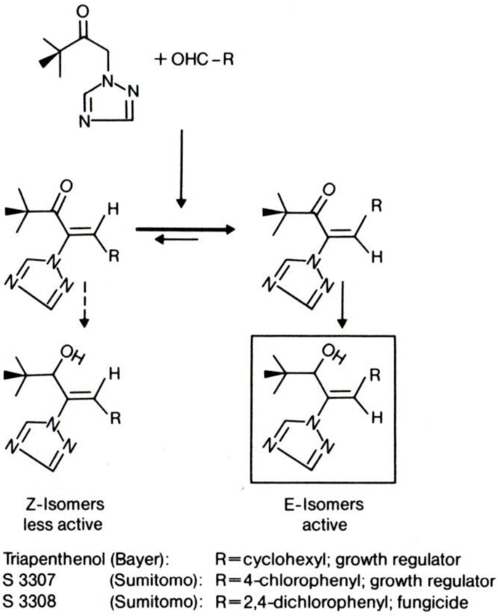

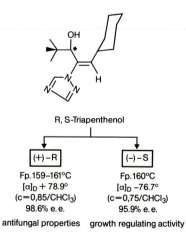

Z-Isomers
less active

E-Isomers
active

Triapenthenol (Bayer):   R=cyclohexyl; growth regulator
S 3307   (Sumitomo): R=4-chlorophenyl; growth regulator
S 3308   (Sumitomo): R=2,4-dichlorophenyl; fungicide

Figure 16.   N-Vinyl-azoles.

R, S-Triapenthenol

(+)–R

Fp.159–161°C
[α]_D + 78.9°
(c=0,85/CHCl$_3$)
98.6% e. e.

antifungal properties

(−)–S

Fp.160°C
[α]_D −76.7°
(c=0,75/CHCl$_3$)
95.9% e. e.

growth regulating activity

Figure 17.   Enantiomers of triapenthenol (RSW 0411).

Figure 18. Plant growth regulating activity of triapenthenol
(RSW 0411) and its (+)-R (KTU 1591; inactive) and (-)-S-enan-
tiomer (KTU 1592; active); barley.

Figure 19. Plant growth regulating activity of triapenthenol
(RSW 0411) and its (+)-R (KTU 1591; inactive) and (-)-S-enan-
tiomer (KTU 1592; active); soybeans.

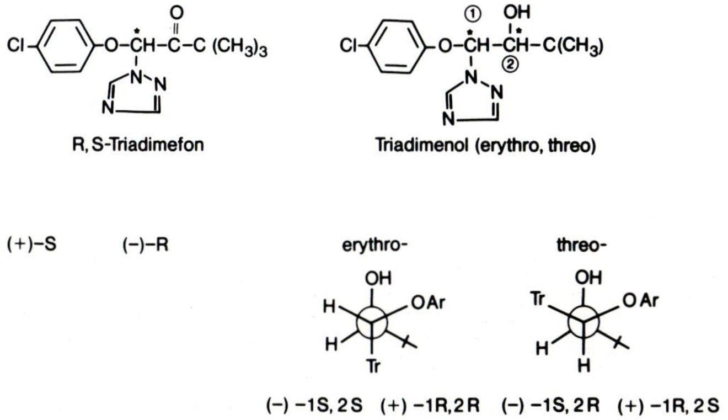

Cl—◯—O–CH–C–C (CH$_3$)$_3$
           ‖
           O
      N–N
      N

R,S-Triadimefon

Cl—◯—O–CH–CH–C(CH$_3$)
              OH ①
              ②
      N–N
      N

Triadimenol (erythro, threo)

(+)-S   (−)-R

erythro-

OH
H— —OAr
H— —
   Tr

threo-

OH
Tr— —OAr
H— —
   H

(−)−1S,2S   (+)−1R,2R   (−)−1S,2R   (+)−1R,2S

Figure 20. Diastereomers and enantiomers of triadimenol. (Reproduced with permission from Ref. 9. Copyright 1983 Pergamon Press.)

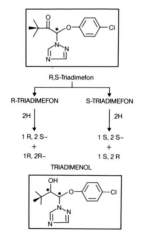

R,S-Triadimefon

R-TRIADIMEFON   S-TRIADIMEFON

2H ↓   ↓ 2H

1R, 2S−   1S, 2S−
   +   +
1R, 2R−   1S, 2R

TRIADIMENOL

Figure 21.   Route to the four enantiomers of triadimenol.

Figure 22. Increasing activity of the four enantiomers of bitertanol against bean rust; (+)-erythro <(-)-erythro < (+)-threo < (-)-threo.

molecular stereochemistry of the azole derivatives also plays an important role when optimizing activity.

A common feature of all products from the azole class is that their fungicidal activity is due to interference with the biosynthesis of fungal steroids, which are essential for cell wall construction in many of these organisms. The interference in ergosterol biosynthesis can occur at one or more points in the biosynthetic pathway depending on the nature of the azole.

In spite of the common mode of action, many other factors are significant for the successful use of individual azole compounds either for control of plant diseases or in the treatment of mycoses. Apart from the nature of the infection, the infection pressure, climatic conditions, the uptake of the fungicide by the plant, its transport and distribution within the plant, and plant compatibility are all important criteria in the complex interaction between pathogen, plant and fungicide. In medicine, high activity, good tolerance and optimal pharmacokinetic properties are prerequisites for therapeutic utility.

In this review, I have presented some of the most important members of the azole class. They represent the thousands of compounds synthesized and biologically tested during the past 18 years. The history of the discovery of the azoles and the results achieved to date serve to illustrate the enormous progress made in the past decade in the research and development of organic fungicides.

Since the early days of plant disease control using sulphur and Bordeaux mixture, an extremely high standard of disease control has been achieved with these new organic fungicides. The highly specific mechanisms of action of these modern agrochemicals differentiate them, for example, from the classical fungitoxic metal salts. The broad, multisite mechanism of action of the latter interfered at many points in the physiological processes of organisms and is no longer regarded as desirable for ecological reasons. Furthermore, successful disease control with these new fungicides is achieved with much lower application rates, both an ecological and an economic advantage. Some of the modern azole products can also be used for curative treatment of plant diseases due to the highly systemic properties. This permits more precise use with fewer prophylactic treatments necessary, a further economic advantage. In medicine as well, the azoles have made significant progress possible in the treatment of human fungal infections.

It seems convincing that the full chemical and biological potential (Table I) in this group of compounds is by no means exhausted. The ever-increasing number of patent applications and the number of publications in scientific journals on this research area are a clear indication that worldwide interest in such a potent class of compounds is very high indeed.

Table I. General Properties of Azoles

--Wide chemical variability without loss of biological properties
--High efficacy against numerous phytopathogenic fungi
--Systemic activity in some cases
--Protective, curative and eradicative activity
--Low application rates
--High activity against fungi pathogenic to humans (dermatophytes,
  yeasts, biphasic fungi, molds)
--Activity against gram-positive bacteria
--Plant growth regulatory activity

Leading Edge of Plant Disease Control

Even today, there are many plant diseases which either cannot be
controlled or where the level of control is inadequate. For these
diseases other methods such as cultural measures and breeding for
resistance have not helped. Considerable losses must still be
accepted, particularly by diseases that attack plants at the root or
base of the stem. Foot and crown rot, caused by Cochliobolus sativus
is a major cereal disease causing great losses in Brazil and Canada.
Take-all disease, caused by Gaeumannomyces graminis, can be found
where cereals are regularly grown and is another of these problem
diseases. Enormous losses are inflicted upon many crops by such
fungi as Pythium, Phytophthora, Verticillium, Fusarium and Rhizoctonia which infect and destroy the roots. Particularly great losses
are caused by fungi that grow into the vascular system of the plant
from the roots and induce wilting.
      In field crops, disease control using soil disinfectants is
unthinkable. The costs and an undesirable effect on soil flora and
fauna are prohibitive for such a method of treatment. An agent is
needed that is absorbed by the aerial plant parts and moves in the
phloem basipetally into the roots where it gives the roots therapeutic protection.
      An apoplastic movement has been observed with the currently
available systemic fungicides. This means that the chemical moves
extracellularly and is transported in the xylem in the direction of
the transpiration flow. With such products, the transfer of chemical
from older, sprayed foliage into newly grown plant parts is not
possible. In contrast, compounds that would be transported actively
from cell to cell (symplastic movement) could move in the phloem,
transport from older, treated foliage into newly grown plant parts.
The use of such products would make the spray timing more independent
of the plant growth. Spray intervals could be lengthened and the
total number of treatments reduced.
      The control of bacterial diseases with the currently available
products is very unsatisfactory. The problem has become urgent with
the spread of fire blight (Erwinia amylovora) in Europe, and the
increased incidence of bacterial leaf blight of rice (Xanthomonas
oryzae) and citrus canker (Xanthomonas citri) as well as many other
bacterial diseases. The direct control of virus diseases is cur-

rently regarded as practically unsolved. If the spread of such diseases cannot be prevented by vector control, or by plant resistance, then we can do nothing but watch helplessly as the losses mount. The situation with mycoplasma-like organisms (MLO) is very similar and are gaining in significance as the cause of plant diseases. As you can see, a wide field of unsolved or unsatisfactorily solved problems lies before us.

Progress in basic phytopathological research will certainly point out new methods for plant disease control in the future. We will need to use all the available weapons, from resistance breeding and quarantine measures, to biological, physical and chemical methods, for the prevention and control of plant disease. In this battle, control by chemical products will continue to play a vital role.

Literature Cited

1. M. Plempel, K. Bartmann, K. H. Büchel and E. Regel Deutsche Med. Wochenschrift 94, 1356 – 1364 (1969).
2. E. F. Godefroi, J. Heeres, J. Van Cutsem and P. A. J. Janssen J. Med. Chem. 12, 784 – 791 (1969).
3. K. H. Büchel, W. Draber, E. Regel and M. Plempel Drugs made in Germany, 15, 208 – 209 (1972).
4. K. H. Büchel, W. Meiser, W. Krämer and F. Grewe VIII. Intern. Plant Prot. Congr., Moscow, Section I, 111 – 118 (1975).
5. H. Kaspers, F. Grewe, W. Brandes, H. Scheinpflug and K. H. Büchel VIII. Intern. Plant Prot. Congr., Moscow, Section II, 398 – 401 (1975).
6. K. H. Büchel, J. Pest. Sci., special issue, 576 – 582 (1977).
7. K. H. Büchel, W. Krämer, W. Meiser, W. Brandes, P. E. Frohberger and H. Kaspers, IX. Intern. Congr. of Plant Prot., Washington D.C., Abstract 475 (1979).
8. W. Krämer, K. H. Büchel, W. Meiser, W. Brandes, H. Kaspers and H. Scheinpflug, Advances in Pesticide Chemistry, Vol. 2, 274 – 279, Pergamon Press, Oxford (1978).
9. G. Jäger, Pesticide Chemistry: Human Welfare and the Environment, Vol. 1, 55 – 65, Pergamon Press, Oxford (1983).
10. K. H. Büchel and M. Plempel, Chronicles of Drug Discovery, Vol. 2, 235 – 269, John Wiley & Sons, Inc., New York (1983).
11. H. A. Staab, Angew. Chem. 74, 407 (1962).

RECEIVED October 1, 1985

Biochemical Mode of Action of Fungicides
Ergosterol Biosynthesis Inhibitors

Dieter Berg

Bayer AG, Agrochemical Division, Research Biochemistry, Plant Protection Center
Monheim, D5090 Leverkusen, Federal Republic of Germany

One or two decades ago the biochemical mode of action of a fungicide
normally was not known. A great part of this was due to the fact
that most of the fungicides were "multi-site effectors" and thus not
accessible to relevant biochemical studies. The situation changed
drastically with the finding of the so-called "single-site effectors"
which could be studied mechanistically. Biochemistry then became an
important assistance for the chemist during optimization of efficacy
within a defined chemical group. By now a tremendous number of different mechanisms have been described. A crude classification of the
modes of action leads to three groups of fungicides: 1) those which
inhibit energy production by blocking SH-groups, the glycolysis/ci-
trate cycle, or the respiratory chain, 2) those that inhibit biosyn-
theses of proteins, nucleic acids, cells walls, and membrane lipids,
or interfere with mitosis, and 3) those which induce indirect effects
which change host/pathogen interactions. An example of this last
group is the induction of phytoalexin production by dichlorodimethyl-
cyclopropane-carboxylic acid (1).

There are numerous examples of the inhibition of biosynthesis by
fungicides. Some fungicidal secondary metabolites, like cycloheximide and blasticidine, interfere with synthesis of peptide bonds at
the ribosomal site (2). Another, kasugamycine, influences aminoacyl-
t-RNA/ribosome interactions (3). Finally, another mechanism inhibiting protein biosynthesis is realized on the DNA/RNA- level by the

0097-6156/86/0304-0025$08.00/0
© 1986 American Chemical Society

acylanilides. For example, metalaxyl interferes with RNA- polymerase (4).

   The first group of systemic fungicides, the benzimidazoles, are inhibitors of mitosis by interference with tubuline polymerization (5). Thus, they prevent an arrangement of the spindle apparatus. The tubuline-benzimidazole-interactions have been studied in detail (6). It is known that carbendazim, for example, after entering the nucleus, specifically binds to the β -subunit of tubuline and by this inhibits the dimerization of the α and β -subunits to a functional tubuline unit. Resistant strains possess altered β -subunits with a decreased affinity for benzimidazoles (7). The modes of action of other inhibitors of mitosis, like dicarboximides, aromatic hydrocarbons, or dithiocarbamates, have not yet been precisely described on a molecular basis.

   Another group of fungicides interferes with cell wall formation. Two examples of this group include: 1) the polyoxines, which prevent chitin formation by competitively inhibiting chitin-synthase, the final enzyme involved in chitin biosynthesis, and 2) melanine biosynthesis inhibitors, especially in Pycularia oryzae. Examples of this type of fungicide are Tricycloazole and Lilolidone, compounds that interfere with pentaketide synthesis (8).

   The next large group includes compounds inhibiting biosynthesis of membrane lipids. The modes of action of sterol synthesis inhibitors will be discussed in detail later, but the mode of action of Kitazin, Isoprothiolane and Edifenphos has also been studied intensively. It could be shown that the S-adenosylmethionine dependent methylation of phosphatidyl-ethanolamines to the corresponding lecithins is affected by these compounds (9). Membranes cannot only be disturbed by rather specific mechanisms but by general actions as well. For example, dodine damages membranes by a detergent-like effect. An example of a different mechanism is the inhibition of adenosine-deaminase by the pyrimidine derivatives ethirimol and dimethyrimol (10).

   In the discussion of the mode of action of ergosterol biosynthesis inhibitors, the question of the function of sterols in membranes has to be asked. A model experiment, conducted by Ladbrooke et al. (11), demonstrates that phospholipid phases are physically stabilized by addition of a sterol, in this case cholesterol. In fungi, cholesterol is replaced by ergosterol, but both sterols are "quasiplanar" and thus are able to function as membrane components. If one simulates a deficiency of a planar sterol, it can be shown by differential calorimetry measurements (Figure 1) that with dispersions of phosphatidylcholine/cholesterol mixtures in water, phase transitions between different lipid phases are induced (11). If, in the case of sterol deficiency, the phase transition temperature is passed between the "quasi-crystalline" and liquid phase of phosphatidylcholines, an energy consuming reaction is observed. This is a quantifiable indication of drastic changes in lipid structure.

Such changes in membrane structure do not only induce drastic changes in the physical stability of membranes, but also affect the specific activities of membrane-bound enzymes. This can be demonstrated with the example of membrane ATP-ase activity of Mycoplasma mycoides (Figure 2). When the enzyme activity is plotted against temperature, the wild-type shows no phase transition in the Arrhenius-diagram. However, in the case of a sterol deficient mutant, the specific activity of the membrane ATP-ase changes at 18°C (12). This clearly indicates that a temperature-dependent change of the membrane conformation causes a temperature-induced change in the specific activity of membrane-bound enzymes.

If sterol content and conformation are so important for membrane stability, we should study the biosynthesis of sterols (Figure 3). The first enzyme in terpenoid biosynthesis is the 3-Hydroxy-3-Methyl-Glutaryl-Coenzyme A-reductase (HMG-CoA-reductase) that catalyzes the synthesis of mevalonate. Two phosphorylations and decarboxylation of mevalonate lead to isopentenylpyrophosphate, the basic C_5-unit in sterol synthesis. Isopentenylpyrophosphate reacts with its isomer, the dimethylallyl-pyrophosphate, in a head/tail-reaction to geranyl-pyrophosphate; reaction with another C_5-unit leads to farnesyl-pyrophosphate, that dimerizes in a tail/tail-reaction to squalene. After expoxidation of its $\triangle^2$-double bond, squalene cyclizes to lano-

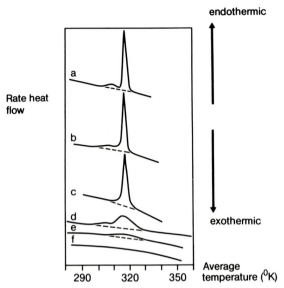

Figure 1. Differential scanning calorimetry curves for 50 wt%
dispersions of 1,2-dipalmitoylphosphatidyl-choline-cholesterol
mixtures in water containing (a) 0; (b) 5.0; (c) 12.5; (d) 20.0;
(e) 32.0; and (f) 50.0 mol% cholesterol. (Reproduced with permission from Ref. 11. Copyright 1968 Elsevier Science Publishing
Company, Inc.)

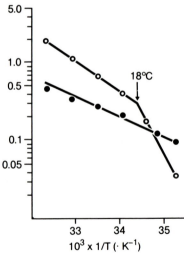

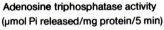

Figure 2. Arrhenius plots of membrane ATP-ase activity of native
(-•-) and sterol deficient (-o-) Mycoplasma mycoides var.
mycoides. (Reproduced with permission from Ref. 12. Copyright
1973 Elsevier Science Publishing Company, Inc.)

sterol, the first sterol but the last common intermediate of both cholesterol and ergosterol synthesis.

In ergosterol biosynthesis, side chain alkylation of lanosterol normally takes place to build 24-methylenedihydrolanosterol, which itself is then the substrate for demethylation reactions at C_{14} and C_4. The C_{14}-demethylation has been studied in detail. It is an oxidative demethylation catalyzed by a cytochrome P_{450}-system. The first step involved in this reaction is the hydroxylation of the C_{14}-methyl-group to form the C_{14}-hydroxymethyl derivative. A second hydroxylation and loss of water lead to the C_{14}-formyl intermediate, which is hydroxylized a third time to form the corresponding carboxylic acid. Decarboxylation does not directly take place, but proceeds instead by abstraction of a proton from C_{15}, followed by elimination and formation of a $\triangle^{14}$-double bond. The NADPH-dependent reduction of the $\triangle^{14}$-double bond finishes the demethylation reaction. Subsequently, demethylation at C_4 has to take place twice, followed by a dehydrogenation reaction in $\triangle^5$-position and isomerization from $\triangle^8$ to $\triangle^7$ and $\triangle^{24(28)}$ to $\triangle^{22}$, respectively.

In research for new fungicides and antimycotics, one has to look for a concept for pathogen-specific inhibitors. This means that inhibition of sterol synthesis should not take place at any common step; the aim is to only inhibit pathogen-specific steps in biosynthesis. The reason for this is to minimize the risk of human toxicity. If one compares the biosynthesis of mammalian cholesterol to that of ergosterol, the main sterol of pathogenic fungi, it becomes obvious that there are at least four pathogen-specific steps to be inhibited.

The first pathogen-specific reaction is the S-adenosylmethionine-dependent side chain aklylation of lanosterol. This is pathogen specific since in cholesterol synthesis, a side chain alkylation does not take place. Secondly, the demethylation reactions at C_{14}- and C_4-positions of 24-methylene-dihydrolanosterol are pathogen-specific as well. In mammals demethylation reactions take place, but the substrate is not side chain alkylated, so the corresponding enzyme should possess different binding sites for the different substrates.

A third pathogen-specific step in ergosterol synthesis is the dehydrogenation in Δ^{22}-position of the side chain. The fourth interesting target is the Δ^8 to Δ^7- isomerization reaction or the
Δ^7-dehydrogenation. These four reactions either do not take place
in mammals, or utilize different substrates.

Another target, that at first seems to be unfavorable since it
is principally common for all organisms, is the enzyme HMG-CoA-reductase which is the regulatory enzyme in terpenoid biosynthesis. Results from trials with naturally produced inhibitors for that enzyme,
such as Compactine and Mevinoline, indicate that these compounds are
able to lower the cholesterol content in mammals, but not markedly
depress sterol synthesis in fungi (13).

There are at least four distinct chemical groups of ergosterol
biosynthesis inhibitors (EBI's). The largest group, in terms of
number of commercial compounds, is the triazole derivatives (Figure
4). The first commercial compound was triadimefon, that can be
chemically reduced to yield triadimenol, which is mainly used as a
seed dressing agent. The main indication of both of these systemic
compounds is for powdery mildew control in cereals. The substitution
of the chlorine atom in triadimenol by a phenyl substituent led to
the synthesis of bitertanol. Bitertanol is mainly active against
rust and scab in tree fruit. Analogous to these phenoxy-derivatives,
the benzyl compound diclobutrazole has been developed. The two
ketals, propicanozole and etaconazole, show spectra of biological
activities that are slightly different from those of triadimefon,
triadimenol, or dichlobutrazole. Topas does not contain the ketal
partial structure, but one still might consider it a propiconazole
analogue. Fluotrimazole is significantly distinct from the other
compounds, but has remarkable homology to clotrimazole, an antimycotic azole.

Apart from the triazole derivatives, a small group of imidazoles
is active against plant pathogens (Figure 5). The most important
compounds in this group are imazalil and prochloraz. However, most
of the important antimycotic EBI's originate from the imidazole series. Two of these antimycotics are bifonazole and clotrimazole,
which resemble fluotrimazole. These two antimycotics will be dis-

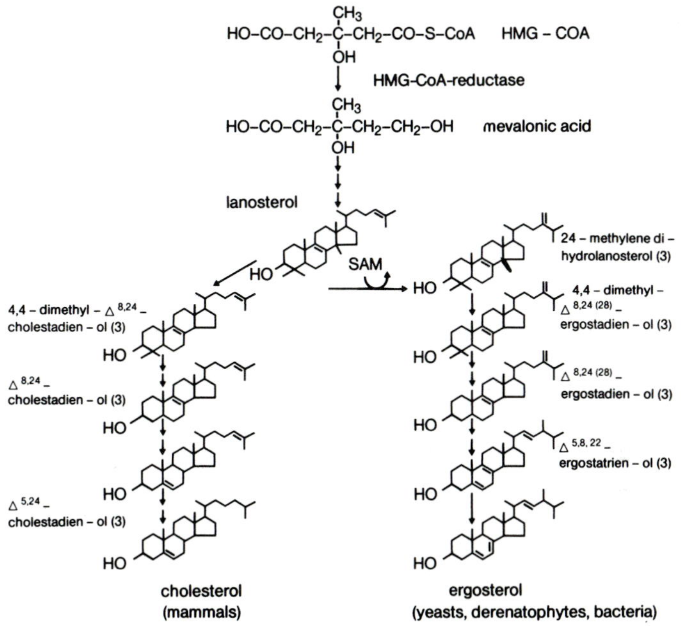

Figure 3.   Strategies for inhibition of ergosterol biosynthesis.

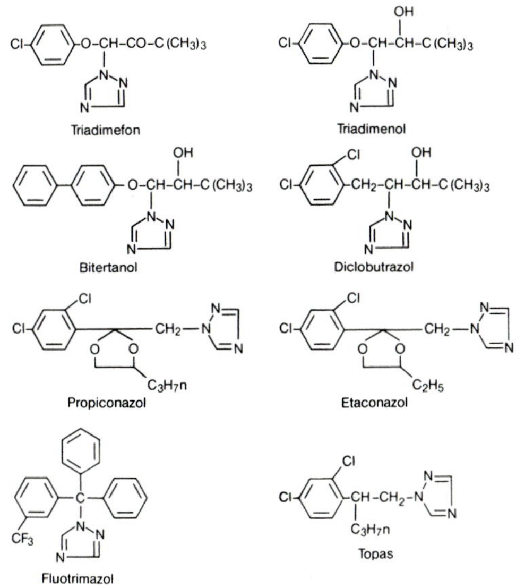

Figure 4.   Triazolyl ergosterol biosynthesis inhibitors.

cussed later with respect to their ability to inhibit HMG-CoA-reductase in dermatophytes.

The third group of EBI's are the pyrimidine derivatives nuarimol, fenarimol, and triarimol (Figure 6). All of these compounds are closely related chemically. The commonality they have with the triazoles and imidazoles is the nitrogen heteroatom in the 3-position from the central carbon. The fourth most interesting group of EBI's are the morpholines which are represented by just two compounds, tridemorph and fenpropimorph (Figure 7).

When we started to look for the mode of action of different EBI's, we decided to use several pathogens as test organisms. These were Pyricularia oryzae, Botrytis cinerea, Cercospora musae, Fusarium nivale, and Drechslera teres. To get rapid information about the possible antimycotic efficacy of a test compound, we also used a non-pathogenic yeast, Saccharomycopsis lipolytica.

In the next part of our research on EBI-fungicides, we restricted ourselves to Pyricularia oryzae since from our point of view the in vitro results with that organism are representative and the test procedure is rather simple. The test chemical is applied in a suitable concentration to the culture medium which is then inoculated from an untreated pre-culture. After a 24-hour fermentation, the cells are separated from the culture filtrate by centrifugation, resuspended in a chloroform/methanol-mixture and homogenized using an ultraturrax treatment. After this extraction procedure, the sterol-conjugates are split to free sterols by a potassium-hydroxide treatment. Adsorption of the sterols to a Sep-pak column and step-wise desorption leads to a sterol fraction which can be analyzed directly by gas chromatography on a SE-30 capillary column.

An example for a GC-analysis of isolated sterols from P. oryzae is shown in Figure 8. An elution diagram of an untreated control is compared to the sterol analysis after application of 10 ppm triadimenol. It can easily be seen that in the region where the sterols are eluted (framed area), the pattern becomes totally different. The elution index is the first criterion for the chemical nature of an accumulating sterol. However, structure elucidation has been per-

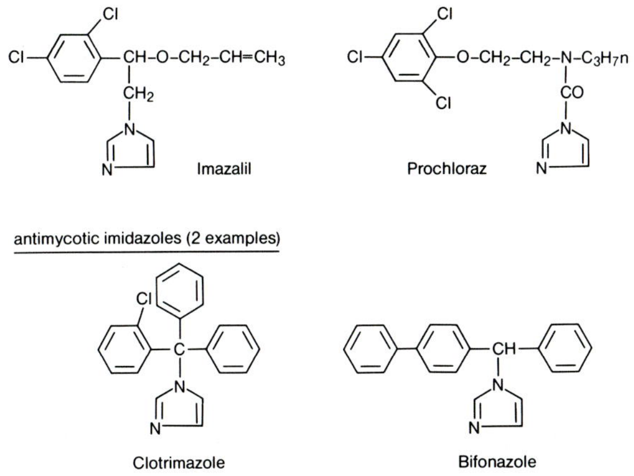

Imazalil

Prochloraz

antimycotic imidazoles (2 examples)

Clotrimazole

Bifonazole

Figure 5. Imidazolyl ergosterol biosynthesis inhibitors.

Nuarimol

Fenarimol

Triarimol

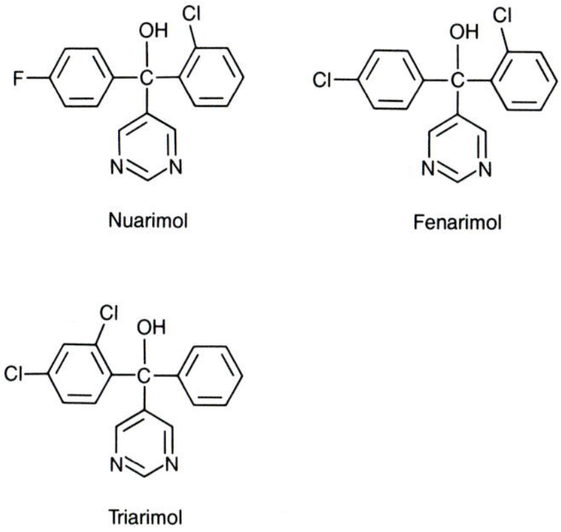

Figure 6. Pyrimidine ergosterol biosynthesis inhibitors.

tridemorph

fenpropimorph

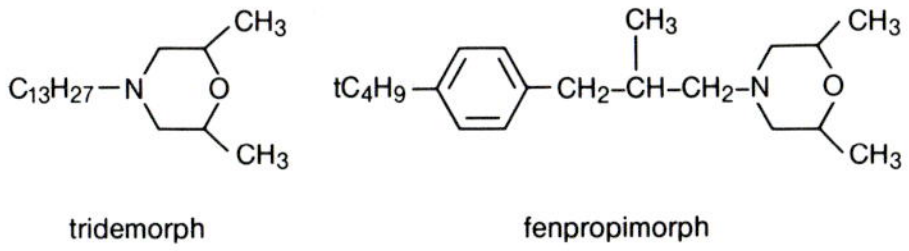

Figure 7. Morpholine ergosterol biosynthesis inhibitors.

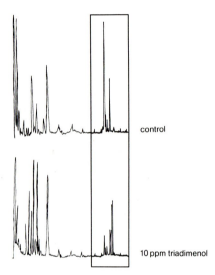

Figure 8. Gas chromatography of sterols from <u>Pyricularia</u> <u>oryzae</u> treated with triadimenol.

formed by GC/MS-coupling experiments; the mass-spectra being compared to those of authentic materials. Quantification of the accumulation was performed by the 100% method with the assumption that the FID-response is the same for all sterols.

The quantity and distribution of the sterol structures in P. oryzae after the fungus had been treated with 10 ppm triadimenol are listed in Figure 9. The untreated control contains, as expected, a large amount of ergosterol as the main membrane component. Besides minor concentrations of $\triangle^5$-ergostenol and $\triangle^{5,22}$-ergostadiene-ol, Pyricularia contains about 15% $\triangle^5$-stigmastenol, but the content of the latter compound is not affected by application of triadimenol. As a consequence of treatment with 10 ppm triadimenol, we observed a 4- fold decrease in ergosterol content and an accumulation of its precursor 24-methylenedihydrolanosterol, a nonplanar compound due to the three methyl groups at C_4 and C_{14} and lack of the $\triangle^5$-double bond. Therefore, the accumulating sterol is not able to function properly as a membrane component.

The site at which triadimenol inhibits ergosterol biosynthesis is illustrated in Figure 10 and the accumulating sterol is framed. This result clearly indicates that triadimenol inhibits the cytochrome P_{450}-dependent oxidative removal of the C_{14}-methyl group of 24-methylenedihydrolanosterol, which is a pathogen-specific precursor of ergosterol.

In the same test system, several other triazoles were examined, including triarimol as an example of the pyrimidine derivatives, and imazalil as a representative of the imidazole series (Figure 11). It is common to all these compounds that 24-methylenedihydrolanosterol accumulates indicating an identical primary mode of action. In the case of triarimol, an additional accumulation of $\triangle^{5,22}$-ergostadiene-ol can be observed, indicating that in Pyricularia oryzae $\triangle^7$-dehydrogenation is also affected by triarimol. These data do not necessarily correspond with biological efficacy since most of these compounds are used to control powdery mildews, which are obligate parasites and are not easily accessible to in vitro studies. Another important condition cannot be fulfilled in such in vitro

sterol [%]	control	triadimenol [10 ppm]
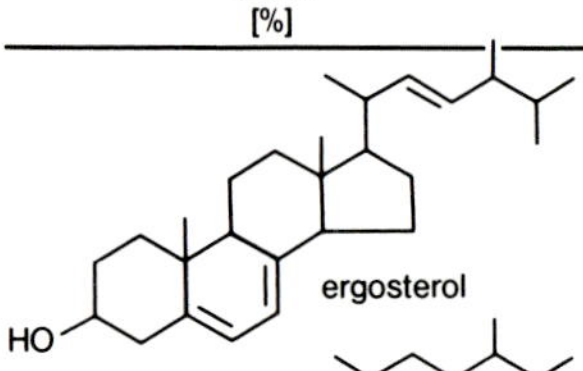 ergosterol	74.7	18.0
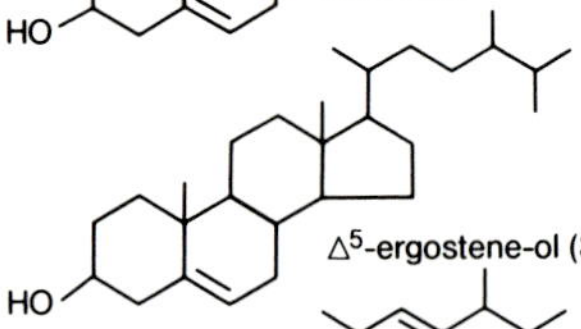 Δ^5-ergostene-ol (3)	5.5	4.2
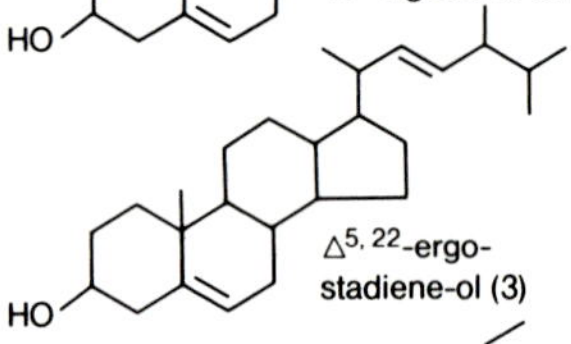 $\Delta^{5,22}$-ergo-stadiene-ol (3)	4.4	7.9
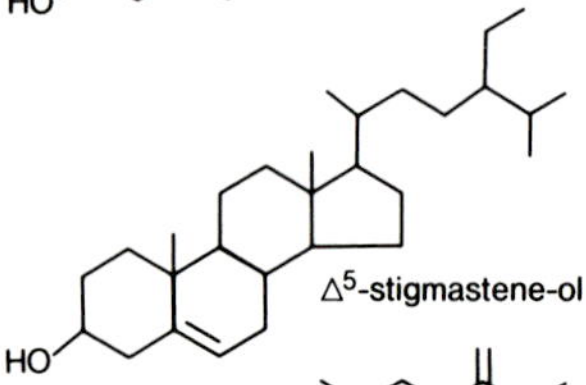 Δ^5-stigmastene-ol (3)	15.4	15.7
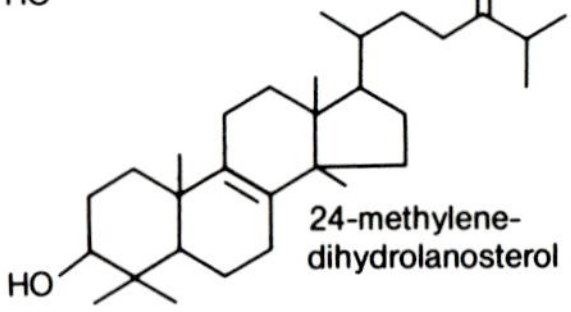 24-methylene-dihydrolanosterol	<0.02	54.2

Figure 9. Distribution of sterols in Pyricularia oryzae after application of triadimenol.

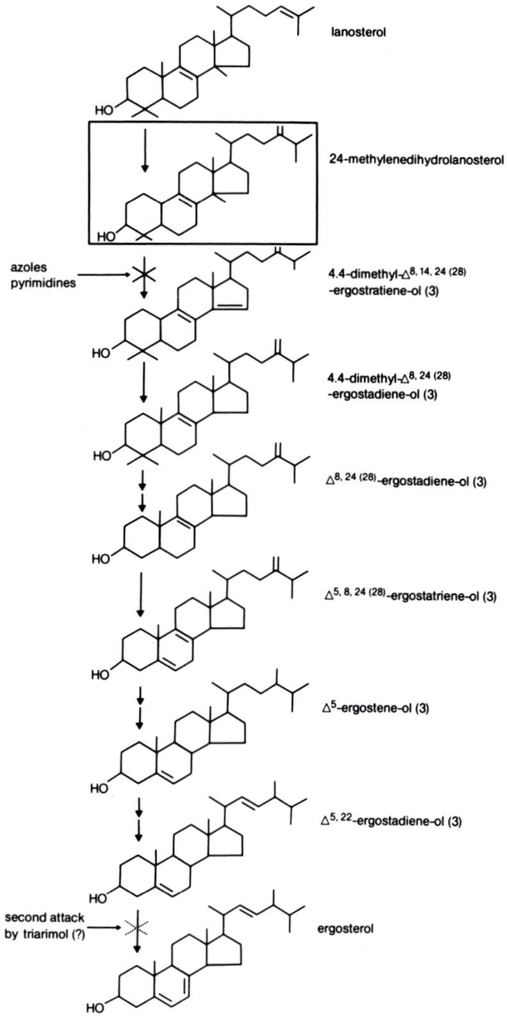

Figure 10. Biosynthetic pathway of ergosterol in Pyricularia oryzae indicating the site of inhibition by azoles and pyrimidines.

sterol	control	triadimefon [10 ppm]	bitertanol [10 ppm]	diclobutrazole [10 ppm]	propiconazole [10 ppm]	etaconazole [10 ppm]	fluotrimazole [10 ppm]	topas [10 ppm]	triarimol [10 ppm]	imazalil [10 ppm]
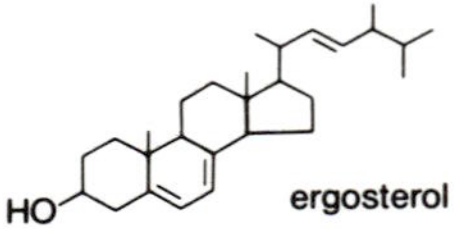 ergosterol	74.7	43.0	10.0	6.1	9.2	14.9	8.0	24.1	14.6	8.7
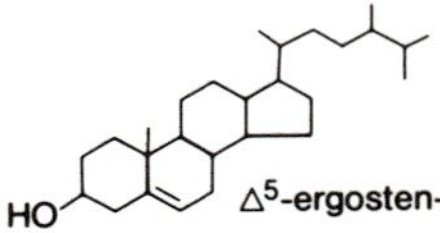 Δ^5-ergosten-ol (3)	5.5	3.8	6.0	4.8	3.4	5.2	5.9	5.6	4.5	6.7
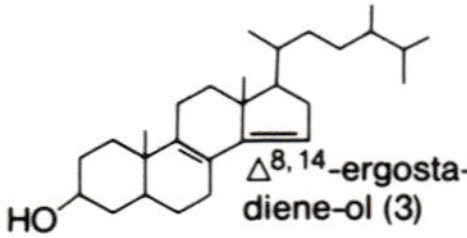 $\Delta^{8,14}$-ergosta-diene-ol (3)	<0.02	<0.02	<0.02	<0.02	<0.02	<0.02	<0.02	<0.02	<0.02	5.6
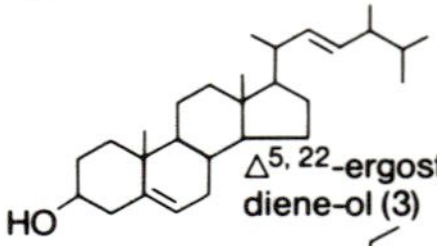 $\Delta^{5,22}$-ergosta-diene-ol (3)	4.4	6.4	6.2	10.9	4.7	15.3	4.5	12.6	32.2	1.2
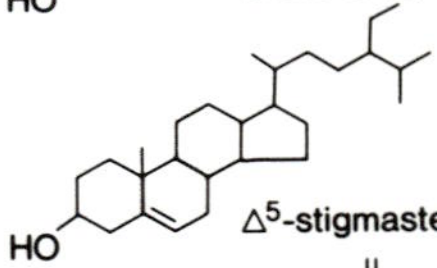 Δ^5-stigmastenol	15.4	11.2	22.0	17.8	18.3	14.7	20.4	17.0	13.8	24.0
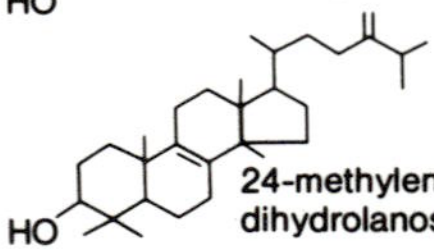 24-methylene-dihydrolanosterol	<0.02	35.6	55.8	60.4	64.4	49.9	61.2	40.7	34.9	53.8

Figure 11. Distribution of sterols in Pyricularia oryzae after application of different azoles and pyrimidines.

studies, namely, a possible metabolic activation during translocation in the plant. For example, triadimefon showed rather weak activity in in vitro studies, but a majority of its efficacy is due to the more active derivative triadimenol, formed by the reduction of the carbonyl group by plant enzymes.

Analogous studies have been performed with the morpholine derivatives fenpropimorph and tridemorph. Starting with fenpropimorph, upon treatment of Pyricularia oryzae, one easily observes a sterol pattern that is quite different from that after application of azoles or pyrimidines (Figure 12). The most characteristic peak of 24-methylenedihydrolanosterol cannot be observed in large amounts. Instead of this precursor, another sterol is accumulated, with a slightly longer elution time than ergosterol, indicating it to be a smaller molecule compared to 24-methylenedihydrolanosterol that does not bear the three methyl groups.

After incubation of Pyricularia oryzae with tridemorph, an analogous picture on the basis of FID-detection is observed (Figure 13). In both cases we looked for the chemical nature of the accumulating sterol using GC/MS-coupling. Surprisingly, the sterols that accumulated after treatment with fenpropimorph and tridemorph were not identical on the basis of mass spectra (Figure 14). In the case of tridemorph, a sterol with the mole peak of 398 accumulates that shows no absorption of a conjugated diene in the UV-region. The mass spectrum is identical to that of authentic $\triangle^{5,8}$-ergostadiene-ol. However, after treatment with fenpropimorph, a sterol accumulates that in all spectroscopic properties is identical to $\triangle^{8,14}$-ergostadiene-ol (ignosterol) that was isolated previously from Ustilago maydis by Kerkenaar (14). What do these results mean with respect to the modes of actions of the morpholines?

The biosynthetic pathway of ergosterol is shown in Figure 15 to explain the steps of inhibition by morpholines; the accumulating sterols are framed. Inhibition of the $\triangle^{14}$-reduction by fenpropimorph should lead to an accumulation of 4,4-dimethyl- $\triangle^{8,14,24(28)}$-ergostatriene-ol. However, C_4-demethylation and side chain hydrogenation are obviously able to occur with the accumulating sterol as substrate

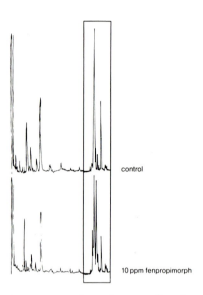

Figure 12. Gas chromatography of sterols from Pyricularia oryzae treated with fenpropimorph.

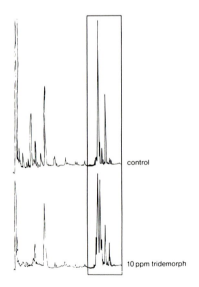

Figure 13. Gas chromatography of sterols from Pyricularia oryzae treated with tridemorph.

sterol [%]	control	tridemorph [10 ppm]	fenpropimorph [10 ppm]
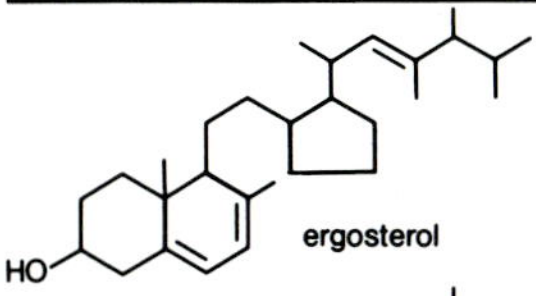 ergosterol	74.7	43.0	47.8
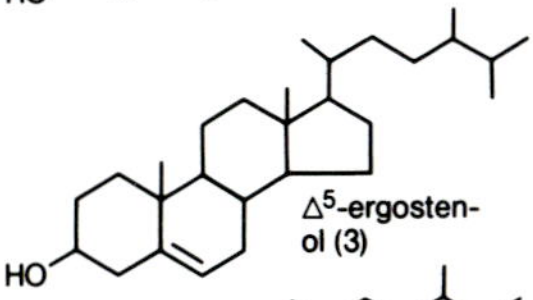 Δ^5-ergosten-ol (3)	5.5	<0.02	<0.02
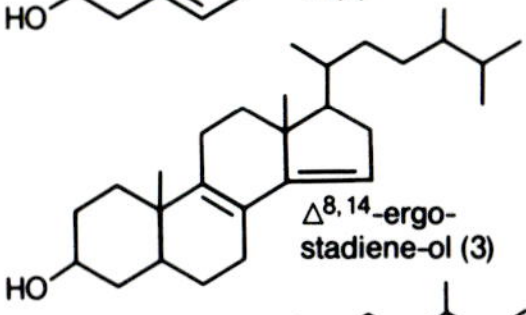 $\Delta^{8,14}$-ergo-stadiene-ol (3)	<0.02	<0.02	32.3
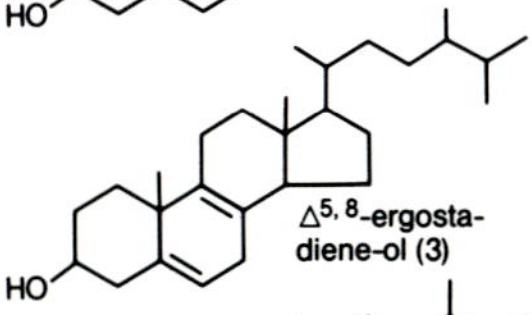 $\Delta^{5,8}$-ergosta-diene-ol (3)	<0.02	31.8	<0.02
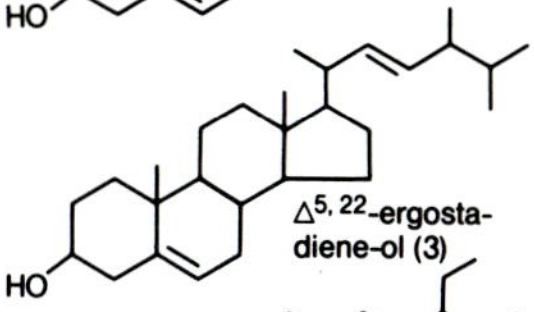 $\Delta^{5,22}$-ergosta-diene-ol (3)	4.4	9.1	6.6
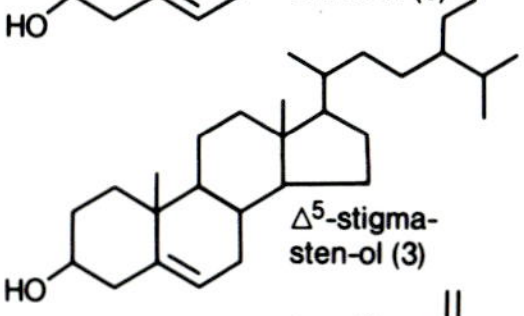 Δ^5-stigma-sten-ol (3)	15.4	11.1	1.6
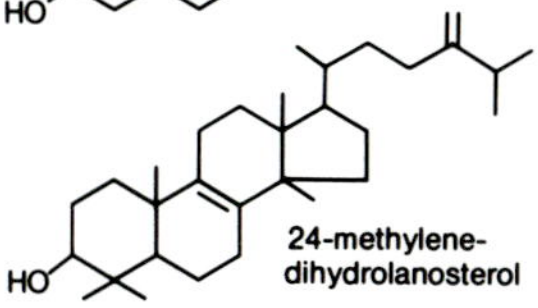 24-methylene-dihydrolanosterol	<0.02	5.0	11.7

Figure 14. Distribution of sterols in Pyricularia oryzae after application of tridemorph and fenpropimorph.

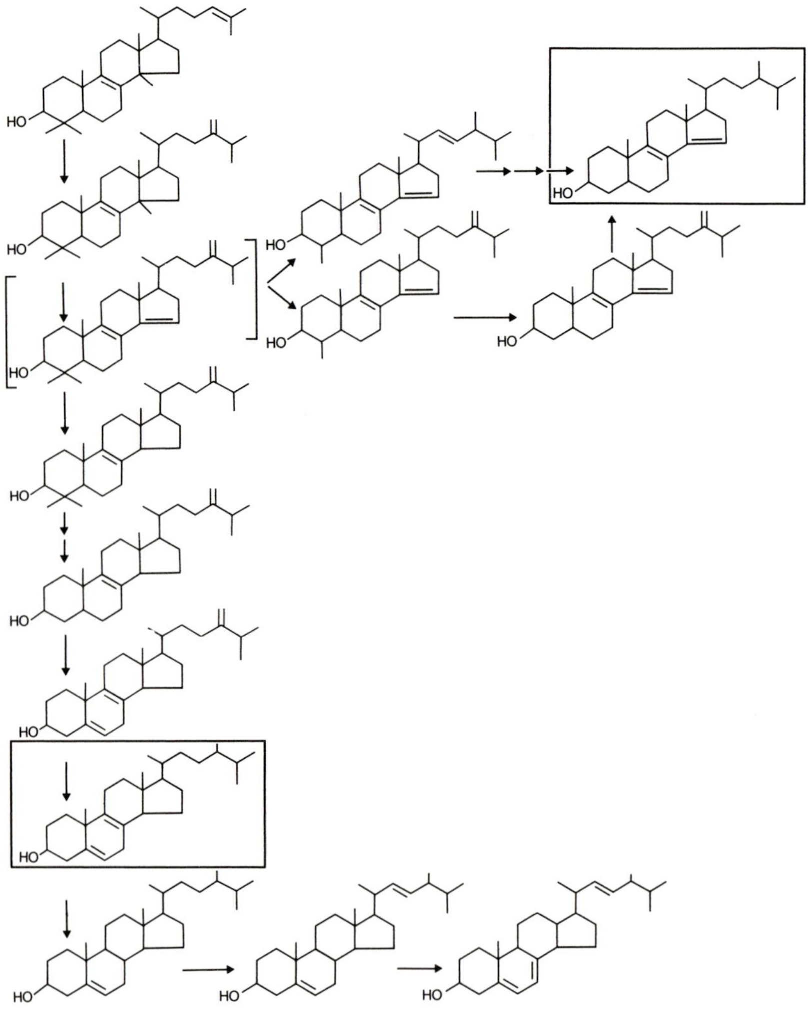

Figure 15. Biosynthetic pathway of ergosterol in _Pyricularia oryzae_ indicating the sites of inhibiton by tridemorph and fenpropimorph.

so that ignosterol, $\triangle^{8,14}$-ergostadiene-ol, is subsequently synthesized. This is in agreement with the results of Kerkenaar and Leroux (14,15). With tridemorph, $\triangle^{5,8}$-ergostadiene-ol accumulates, indicating an inhibition of the $\triangle^{8}$ to $\triangle^{7}$-isomerization reaction. This is in agreement with the finding of Kato (16) with Botrytis as a test organism, but in contrast to the results of Kerkenaar (14) who postulated an identical mechanism of the two morpholines with $\triangle^{14}$-reduction in Ustilago.

The conclusion is that the $\triangle^{8}$ to $\triangle^{7}$-isomerization reaction starts with $\triangle^{8}$-hydrogenation, so that both morpholines inhibit NADPH-dependent reduction steps. In the case of tridemorph, it may vary from organism to organism whether $\triangle^{8}$ or $\triangle^{14}$-reduction is affected. From these results, one might have the impression that the morpholines show some variation in their modes of action but that the azoles always show identical behavior with respect to mechanistic studies. To demonstrate that this is not the case, we chose the examples of clotrimazole and biofonazole, two antimycotic imidazoles, as here we were able to study the human pathogens directly.

To determine whether our strategies could be applied to all of the human pathogens we studied, the sterol composition and the rate of sterol biosynthesis in untreated organisms have been estimated (Figure 16). All the organisms synthesize ergosterol up to 90 % or more with the exception of Trichophyton mentagrophytes. T. mentagrophytes synthesized mainly ergosterol and $\triangle^{8,24(28)}$-ergostadiene-ol, as well as minor amounts of different sterols. Both major sterols are "quasi-planar" and thus may function as membrane components. Additionally, the occurence of $\triangle^{8,24(28)}$-ergostadiene-ol indicates that in Trichophyton mentagrophytes, $\triangle^{5}$-dehydrogenation should be rate limiting.

When Candida albicans was treated with bifonazole or clotrimazole, the expected accumulation of the ergosterol precursor 24-methylenedihydrolanosterol was observed (Figure 17). This was an indication that both compounds inhibit the cytochrome P_{450}-dependent C_{14}-demethylation. Using 2.5 µg /ml clotrimazole, dihydrolanosterol accumulates, which is known to regulate the HMG-CoA-reductase by a

organisms

sterol [% total sterols]	Candida albicans [48 h]	Candida albicans [Pseudomycel/72h]	Microsporum canis	Trichophyton mentagrophytes [48 h]	Trichophyton mentagrophytes [72 h]	Epidemophyton floccosum	Torulopsis glabrata
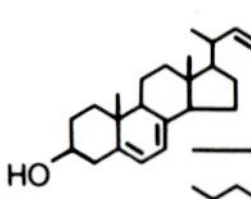 Ergosterol	93,6	83,5	98,6	54,9	47,6	100	89,7
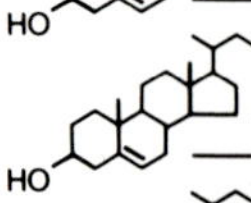 Δ^5-Ergosten-ol (3)	<0,02	<0,02	<0,02	3,9	5,1	<0,02	<0,02
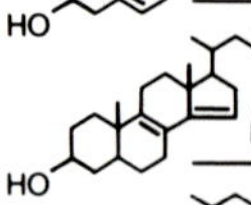 $\Delta^{8,14}$-Ergostadien-ol (3)	<0,02	<0,02	<0,02	<0,02	0,5	<0,02	<0,02
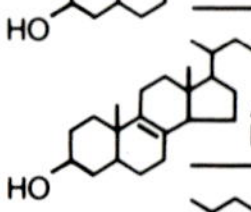 $\Delta^{8,24\,(28)}$-Ergostadien-ol (3)	4,2	<0,02	<0,02	37,4	40,1	<0,02	3,2
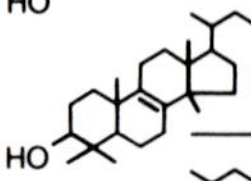 Dihydrolanosterol	2,2	11,6	1,4	<0,02	1,5	<0,02	5,6
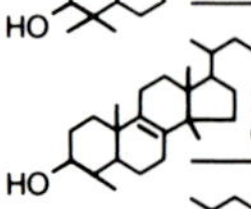 4,14-Dimethyl-$\Delta^{8,24\,(28)}$-ergostadien-ol (3)	<0,02	<0,02	<0,02	1,2	1,6	<0,02	<0,02
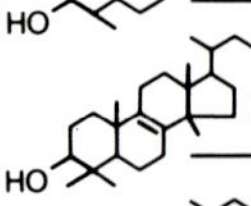 Lanosterol	<0,02	2,4	<0,02	0,9	1,7	<0,02	1,5
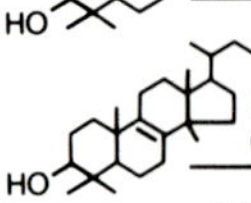 24-Methylen-dihydrolanosterol	<0,02	2,5	<0,02	1,7	1,9	<0,02	<0,02
total sterols/test [µg]	439	212	1500	1026	1185	54	722

Figure 16.   Distribution of sterols in untreated human pathogens.

feed-back control (17), causing a decreased total sterol synthesis. Bifonazole primarily accumulates 24-methylenedihydrolanosterol, but the content of dihydrolanosterol increased at the 5 μg/ml dosage rate. This reflected in the total amount of sterols. Both accumulating sterols are unable to function properly as membrane components.

Since the invasive form of Candida in vaginal mycosis is the pseudo-mycelium, we also looked at this morphologically specialized material. We observed that bifonazole causes an accumulation of dihydrolanosterol exclusively, whereas clotrimazole causes the normal accumulation of 24-methylenedihydrolanosterol, dihydrolanosterol, and lanosterol (Figure 18). However, after bifonazole application, the rate of total sterol synthesis is lowered by a factor of two, a result which will be discussed later.

The results with bifonazole, in contrast to clotrimazole, show that the sterol pattern could not completely explain the rate of sterol biosynthesis. Torulopsis glabrata is another example where both compounds show the expected accumulation of 24-methylenedihydrolanosterol and dihydrolanosterol (Figure 19). The accumulation of dihydrolanosterol, after application of 2.5 μg/ml bifonazole, does not explain why the total sterol content is decreased so markedly. A "feed-back" control of the HMG-CoA-reductase appears an unlikely explanation.

A very dramatic example of this effect is seen in Trichophyton rubrum where clotrimazole, as well as bifonazole, causes no major effects on the sterol pattern (Figure 20). Clotrimazole at 10 μg/ml causes a slight accumulation of methylated sterols, reflecting its biological activity. Bifonazole causes only very small changes, but markedly decreases the total sterol content, resulting in good activity against that organism. This effect of bifonazole has been especially observed in dermatophytes. Therefore when bifonazole was used, there appeared to be an additional inhibition step in the terpenoid biosynthesis between HMG-CoA-reductase and squalene. Since it is known that HMG-CoA-reductase regulates all terpenoid synthesis (17), we looked at this enzyme in a way that excludes the possibility of "feed-back" control.

sterol [% total sterols]	control (EtOH)	Clotrimazole (2,5 µg/ml)	Clotrimazole (5 µg/ml)	Bifonazole (2,5 µg/ml)	Bifonazole (5 µg/ml)
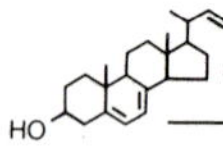 ergosterol	93,6	<0,02	<0,02	10,5	<0,02
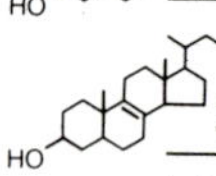 $\Delta^{8,24(28)}$-ergostadien-ol (3)	4,2	<0,02	<0,02	10,0	<0,02
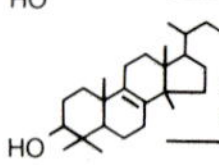 dihydrolanosterol	2,2	40,8	47,6	23,4	43,9
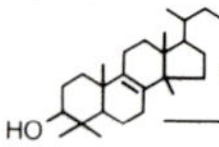 lanosterol	<0,02	<0,02	<0,02	4,5	<0,02
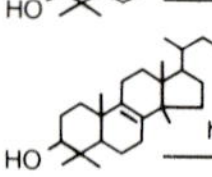 24-methylen-dihydrolanosterol	<0,02	59,2	52,4	51,6	56,1
total sterols/test [µg]	439	86	52	391	127

Figure 17. Distribution of sterols in _Candida_ _albicans_ after incubation with clotrimazole and bifonazole.

sterol [% total sterols]	control (EtOH)	Clotrimazole (0,1µg/ml)	Clotrimazole (0,2µg/ml)	Bifonazole (0,1µg/ml)	Bifonazole (0,2µg/ml)
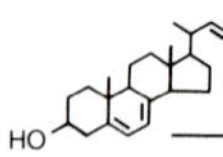 ergosterol	83,5	<0,02	<0,02	82,7	75,2
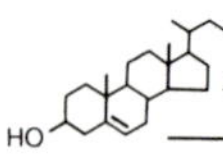 Δ^5-ergosten-ol (3)	<0,02	16,0	18,5	<0,02	<0,02
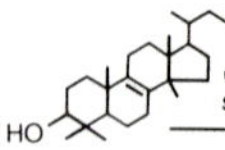 dihydrolanosterol	11,6	35,1	35,1	17,3	24,8
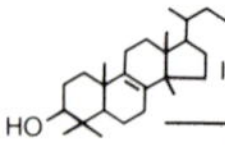 lanosterol	2,4	15,8	19,5	<0,02	<0,02
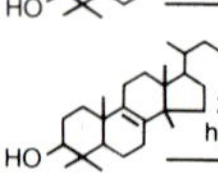 24-methylen-dihydrolanosterol	2,4	33,1	26,9	<0,02	<0,02
total sterols/test [µg]	212	202	204	130	123

Figure 18. Distribution of sterols in pseudomycelium of _Candida_ _albicans_ after incubation with clotrimazole and bifonazole.

sterol [% total sterols]	control [EtOH]	Clotrimazole (1 µg/ml)	Clotrimazole (2,5 µg/ml)	Bifonazole (1 µg/ml)	Bifonazole (2,5 µg/ml)
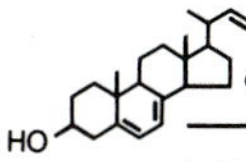 ergosterol	89,7	16,0	8,7	4,6	<0,02
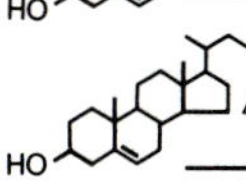 Δ^5-ergosten-ol (3)	<0,02	10,1	12,7	12,4	<0,02
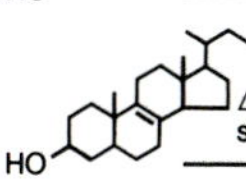 $\Delta^{8,\,24\,(28)}$-ergo-stadien-ol (3)	3,2	<0,02	<0,02	<0,02	<0,02
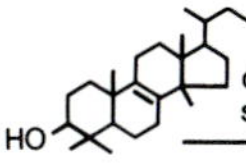 dihydrolano-sterol	5,6	18,7	20,2	20,7	39,0
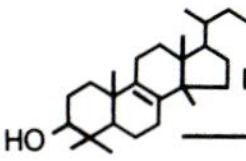 lanosterol	1,5	55,2	58,4	62,3	61,0
total sterols/test [µg]	722	727	431	612	116

Figure 19. Distribution of sterols in <u>Torulopsis</u> <u>glabrata</u> after incubation with clotrimazole and bifonazole.

sterol (% total sterols)	control	Clotrimazole		Bifonazole	
		(5 ng/ml)	(10 ng/ml)	(5 ng/ml)	(10 ng/ml)
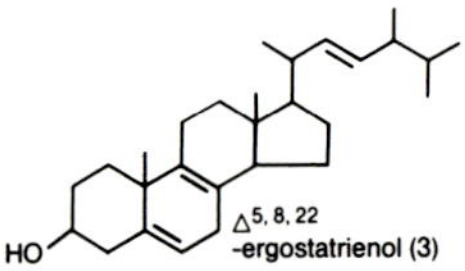 Δ5, 8, 22 -ergostatrienol (3)	5.1	8.3	7.7	6.5	6.5
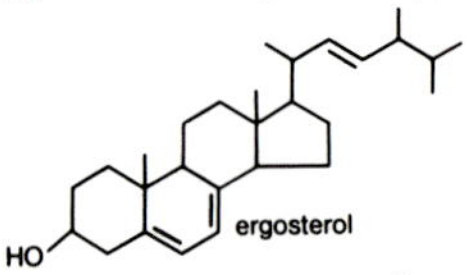 ergosterol	89.6	81.0	74.2	78.2	76.8
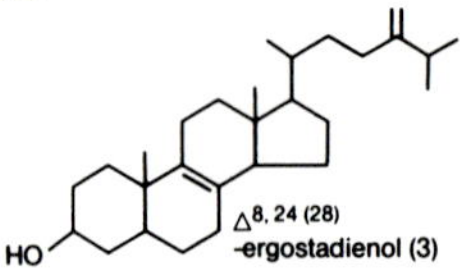 Δ8, 24 (28) -ergostadienol (3)	5.3	4.8	9.7	2.4	2.8
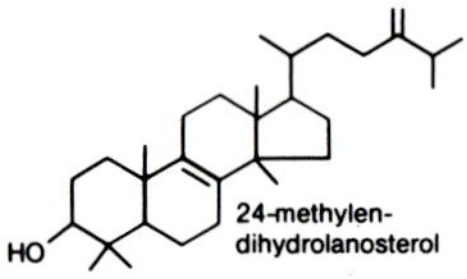 24-methylen-dihydrolanosterol	< 0.02	5.9	8.4	13.1	13.9
total sterols/test (µg)	983	815	723	501	373

Figure 20. Distribution of sterols in Trichophyton rubrum after incubation with clotrimazole and bifonazole.

Study of the HMG-CoA-reductase was done by the isolation of microsomal subcellular fractions and subsequently testing the conversion of HMG-CoA to mevalonate. Using <u>Trichophyton mentagrophytes</u> as a representative of dermatophytes, we isolated microsomes by differential centrifugation. Clotrimazole was not able to inhibit this enzyme at up to 2 µg/ml (Figure 21). However, bifonazole strongly inhibits the enzyme. Under our test conditions at greater than 1 µg/ml, only 20% of the enzyme activity remained compared to the control experiment. We therefore concluded that bifonazole, in contrast to clotrimazole, directly inhibited the HMG-CoA-reductase, thus leading to a sequential action that was responsible for not only the fungistatic but also the fungicidal effects of bifonazole (<u>18</u>, <u>19</u>).

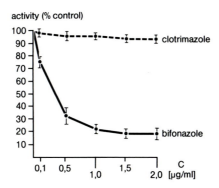

Figure 21.   Inhibition of microsomal HMG-CoA-reductase from <u>Trichophyton</u> <u>mentagrophytes</u> with clotrimazole and bifonazole.

Conclusions

Ergosterol biosynthesis inhibitors differ in their molecular modes of action. This is especially true for the morpholine derivative tridemorph and fenpropimorph. In <u>Pyricularia oryzae</u>, tridemorph inhibits the $\triangle^8$ to $\triangle^7$ -isomerization reaction in ergosterol synthesis, while fenpropimorph prevents the NADPH/H+ -dependent reduction of the $\triangle^{14}$-double bond. Azoles and pyrimidines primarily inhibit the hydroxylation of the C_{14}-methyl group, which is the initial step in the oxidative demethylation reaction. Within this group of inhibitors, differences do occur due to additional effects of various compounds, as was found with the antimycotic imidazoles (clotrimazole and bifonazole). Therefore, each EBI fungicide needs to be studied as an individual so that these additional effects may be observed.

Literature Cited

1. D. Cartwright, P. Langcake, R.J. Pryce, D.P. Leworthy, J.P. Ride, <u>Nature</u> 267 (1977) 511.

2. H.D. Sisler, <u>Ann. Rev. Phytopathol.</u> 7 (1963) 311.

3. A. Okuyama, et al., <u>J. Antikiotic.</u> 28 (1975) 903.

4. L.C. Davidse, A.E. Hofman, G.C.M. Velthuis, <u>Exp. Mycol.</u> 7 (1983) 344.

5. A. Kaars Sijipestijn, in Dekker and S.G. Georgopoulos (ed's): <u>Fungicide Resistance in Crop Protection</u>, Centre for Agricultural Publishing and Documentation, (1982) 32.

6. J.P. Laclette, G. Guerra, C. Zentina, <u>Biochem. Biophys. Res. Commun.</u> 92 (1980) 417.

7. L.C. Davidse, in J. Dekker and S.G. Georgopoulos (ed's): <u>Fungicide Resistance in Crop Protection</u>, Centre for Agricultural Publishing and Documentation (1982) 60.

8. P.M. Wolkow, H.D. Sisler, <u>Phytopathology</u> 72 (1982) 712.

9. O. Kodama, H. Yamada, T. Akatonka, <u>Agr. Biol. Chem.</u> 43 (1979) 1719.

10. D.W. Hollomon, <u>Proc. 1979 Brit. Crop Protect. Conf. Brighton,</u> 251.

11. B.D. Ladbrooke, R.M. Williams, D. Chapman, <u>Biochem. Biophys.</u> <u>Acta</u> 50 (1968) 333.

12. S. Rottem, V.P. Cirillo, B. de Kruyff, M. Shinitsky, S. Razin, <u>Biochem. Biophys. Acta</u> 323 (1973) 509.

13. D. Berg unpublished results.

14. A. Kerkenaar, J.M. von Rossum, J.G. Versluis, J.W. Marsmann, <u>Pestic. Sci.</u> 15 (1984) 177.

15. P. Leroux, M. Gredt, Comptes Rendus des Seances de l'Academie des Sciences, <u>Sciences de la vie, Paris 296</u>, 4 (1983) 191.

16. T. Kato, M. Shoami, Y. Kawase, <u>J. Pestic. Sci.</u> 5 (1980) 69.

17. D. Berg. W. Draber, H. von Hugo, W. Hummel, D. Mayer, <u>Z. Natur-</u> <u>forsch. 36c</u> (1981) 798.

18. M. Plempel, E. Regel, K.H. Büchel, <u>Arzneim. Forsch. 33</u>, 1 (1983) 517.

19. D. Berg, E. Regel, H.E. Harenberg, M. Plempel, <u>Arzneim. Forsch.</u> <u>34</u>, 1 (1984) 139.

RECEIVED October 1, 1985

Uptake and Translocation of Carbon-14-Labeled Fungicides in Cereals
Macro- and Microautoradiographic Studies

Fritz Führ

Institute of Radioagronomy, Nuclear Research Center Jülich GmbH, D-5170 Jülich, Federal Republic in Germany

Insecticides played the dominant role in plant protection until about 1960 when they were displaced in total application by herbicides. Herbicides were especially important in reducing wage-intensive work and resulted in a considerable increase in agricultural crop production. As crops were planted more densely, infestation pressure from fungal diseases also increased and therefore fungicides have become increasingly important. New fungicides like azole compounds, with a wide spectrum of antimycotic and fungicidal activity (1-7) have been introduced and have replaced some of the fungicide standards such as the mercury seed dressing products. These new fungicides used as seed dressings are effective at low dosages and protect the plants from infection by certain seed- and soil-borne plant pathogens as well as against early season infections of powdery mildew and rust fungi (8, 9). Because of improved vigor, the small grain plant develops additional earbearing stalks (10-13). In addition, treatment with these new fungicides during the last phase of grain filling protects the flag leaf and allows the plant to assimilate over a long period of time, so that the genetically determined productivity of the plant can be fully exploited.

0097-6156/86/0304-0053$06.00/0
© 1986 American Chemical Society

Utilization of Radiocarbon ^{14}C.

To improve our use of these new fungicides, detailed studies need to be conducted to determine the leaf or root uptake and transport processes as well as the distribution mechanisms of the new fungicides and/or their metabolic compounds through cells, cell membranes and other plant barriers. Recent studies have concentrated on translocation in the two transport systems of the plant, the xylem and the phloem. In the xylem (lignified cells), water and nutrient transport largely takes place in the direction of the transpiration stream, from the root to the leaf. In the phloem (living transport tissue), certain plant nutrients, sugars, amino acids and plant-specific compounds are distributed throughout the plant but mainly from the leaves or roots into the seeds or other storage organs. The distribution and ultimate plant area protected by a fungicide is determined by the effective transportation in both systems. The possible uptake and translocation mechanisms have been discussed by Edgington and Peterson (14), Crowdy (15) and Buchenauer (16).

Successful analysis of small quantities of the chemical being translocated requires the use of radioactive isotopes such as ^{14}C. The Institute of Radioagronomy has been carrying out studies with ^{14}C-labelled active plant protection substances for 13 years (17, 18). The results indicate that intensive cooperation between plant protection chemists, phytopathologists, phytophysiologists and specialists in the radioisotope techniques is necessary to fully exploit the application possibilities and to interpret the results. The special experimental facilities at the Jülich Nuclear Research Center which include practically oriented field tests supplemented by detailed studies under defined climatic conditions enable practical and relevant results to be obtained (17-19). The aim of this contribution is to provide new insights and information on the system effectiveness and residue behavior of azole fungicides.

The following three active substances were used in this study (Figure 1): [benzene ring-U-^{14}C]triadimenol, [benzene ring-U-^{14}C]-triadimefon and [^{14}C]fluotrimazole. These studies are included as

Triadimefon

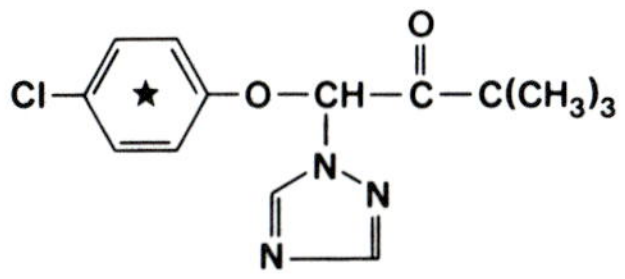

1-(4-Chlorphenoxy)-3,3-dimethyl-1-(1,2,4-triazol-1-yl)-2- butanon

Triadimenol

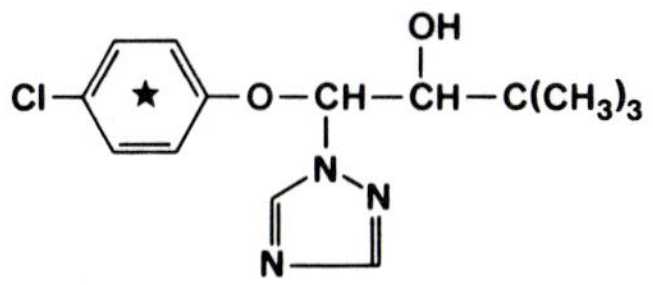

1-(4-Chlorphenoxy)-3,3-dimethyl-1-(1H-1,2,4-triazol-1-yl)-2- butanol

Fluotrimazol

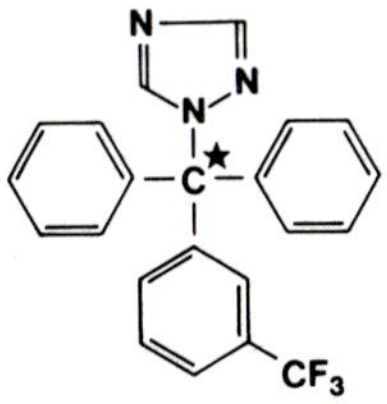

Bis-phenyl-(3-Trifluormethylphenyl)-(1,2,4,Triazolyl)-methan

Figure 1. Fungicidal substances applied and ^{14}C-labeling positions (*).

active components in the commercial preparations Baytan, Bayleton and Persulon, respectively.

Cereal Seed Dressing with Triadimenol

The uptake and distribution of [benzene ring-U-^{14}C]triadimenol applied as a seed dressing in spring barley and spring wheat was investigated under field conditions (19). Lysimeter experiments in small agrarian ecosystems (0.25 - 1 m^2) filled with either topsoil (1 m^2 lysimeter) or with undisturbed soil cores from arable land were used (20, 21). The test soil was a loess loam (parabrown soil, alfisol) which is among the most fertile soil in the whole Federal Republic of Germany (22). Calculation using the specific activity of the labeled fungicide indicated that a seed dressing application of 16 or 12 µg active substance/grain (Table I) corresponded to 177 or 160 g Baytan /100 kg seed (19).

Table I: Radioactivity and Quantity of Active Substance
on the Seed Grain after Dressing with
[benzene ring-U-^{14}C]Triadimenol

	Crop	
	Spring Barley	Spring Wheat
Thousand grain weight (g)	36.2	30.0
Radioactivity/grain* (µ Ci)	1.08	0.83
Active substance/grain* (mg)	0.016	0.012
Active substance/kg seed grain (mg)	442	400
Baytan 25 DS/100 kg seed grain (g)	176.8	160

*Mean values of 50 grains

The distribution of radioactivity in the plant was determined using X-ray films (macroautoradiographs). The radioactive radiation causes blackenings on the film. During the early development up to shoot elongation, a maximum of 7.5% of the radioactivity applied was taken up and translocated into the wheat stalks and leaves with the majority of the translocation occurring between the tenth and twenty-eighth day after sowing. These macroautoradiographs of the wheat plants show that the radioactive substance of the metabolites were translocated into the leaf tips with the transpiration stream (Figure 2).

By means of special film, the radiocarbon of triadimenol in the plant cells can also be detected in thin tissue sections (8 μm) produced with the freezing microtome. Blackening on the film reflects the position of the radioactive substances in the plant cells (23, 24). These microautoradiographs (Figure 3) show that during the swelling phase of the grains radioactivity migrates into the longitudinal and cross cells of the pericarp. Only a small amount moves through the testa into the aleurone layer, which is largely the protein storage layer in the grain. During the first six days of germination, starch is broken down into sugar and mobilized with the protein from the aleurone layer. According to these microautoradiographs, during the first week of germination the perocarp is apparently an effective barrier against the penetration of the active substance into the interior of the grain.

Triadimenol uptake can occur by contamination of the coleoptiles. The radioactive seed dressing was taken up during the first three days of germination of the wheat seedling (Figure 4a) and could then be found in all the cells of the seedling (19). A clear distribution gradient of radioactivity from the outer to the inner cells of the seedling is thus visible. Thirty days after sowing, radioactivity can be detected in cells of the leaf tips (Figure 4b) indicating the active substance and/or its metabolites has migrated with the transpiration flow after uptake in the transport tissue (xylem) followed by translocation into neighboring cell regions.

The dressed grains were sown in the soil at a depth of 3-4 cm. Swelling and germination followed as a result of absorbing water from the ambient soil. Interim results at various points in the development showed that during the swelling phase up to 80% of the active substance only superficially attached to the grain during the dressing, was rapidly transferred into the soil. Special ^{14}C-analyses of

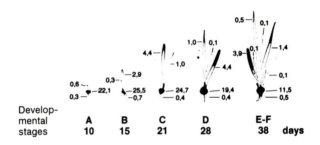

Develop-mental stages	A 10	B 15	C 21	D 28	E-F 38 days

Figure 2. Macroautoradiographs of spring wheat plants in various stages of development after seed dressing with [^{14}C]Baytan. (Reproduced with permission from Ref. 19. Copyright 1982 Pflanzenschutz-Nachrichten Bayer.)

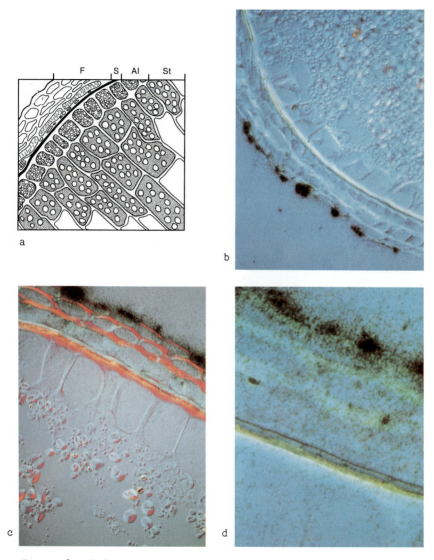

Figure 3. Enlarged sections of microautoradiographs of freezing microtome cross sections through wheat grains after seed dressing with [^{14}C]Baytan (interference phase contrast pictures). (a) Schematic (25) of wheat grain cell structures. Key: F, pericarp; S, testa; AL, aleurone layer; and St, starchy endosperm. (b) Wheat grains immediately after treatment with [^{14}C]Baytan. (c) Wheat grains after 6 days swelling and germination in the soil. (d) Wheat grains after 6 days swelling and germination in the soil (more greatly magnified). (Reproduced with permission from Ref. 19. Copyright 1982 Pflanzenschutz-Nachrichten Bayer.)

Primary leaf (30 days)

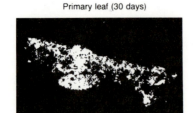

a

Wheat embryo (3 days)

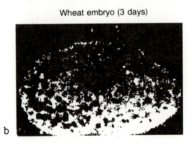

b

Figure 4. Microautoradiographs of freezing microtome cross sections through a spring wheat embryo and the primary leaf tip
after seed dressing with [^{14}C]Baytan (dark-field micrograph).
(a) Cross section through the embryo 3 days after swelling. (b)
Cross section through the primary leaf 30 days after sowing the
dressed seed. (Reproduced with permission from Ref. 19. Copyright 1982 Pflanzenschutz-Nachrichten Bayer.)

soil segments confirmed that, depending on the test soil, a radioactivity distribution gradient is formed around the grain up to a distance of 8 cm. A "dressing area" results around the grain with concentrations of active substance and/or metabolites decreasing outwardly (26).

In recent studies, when non-radioactively dressed winter wheat grains were alternately planted in a series with radioactively dressed grains, radioactivity was detected in the non-treated plants indicating an uptake of ^{14}C-labelled compounds from the "dressing area" via the root. Uptake was about the same order of magnitude as uptake via the grain and seedling (27). Only a combination of radioisotope measurements and chemical analysis will confirm whether this is still the active substance (28). Inoculation with appropriate pathogens is needed to determine whether the concentration of the active substance in the tissue is sufficient to provide protection against these infections.

A conclusion from these studies is that cereal dressing may be a very economical method to prevent fungal plant diseases in the early phase of development. New systemically active substances of the triazole type control the pathogenic fungi in the seed as well as on this surface. Uptake of the active substance by the seedling and its transport into the epigeal parts of the plant results in the protection of the seedling and the primary leaf from both soil- and air-borne fungal pathogens. This detailed information combined with the results of chemical and radiochemical analyses of this type can provide information to aid in the improvement of seed dressing formulations and techniques with the objective of applying only the appropriate amount of active substance and at the location where plant uptake is most probable.

Uptake of Triadimefon via the Leaves

If spraying is carried out during the period of intensive leaf development (e.g. during tillering and especially during development of the stalks in cereals), it is almost impossible to achieve uniform distribution on the plant surfaces. Therefore, information on the internal transport of a new fungicide with the aid of ^{14}C-labelling is important, in order to fully exploit their potential broad spectrum of activity. Only compounds with a systemic effect can combat fungal diseases on the plant as well as eliminate pathogens after infection since the active substance must penetrate into the plant tissue and be transported to non-treated plant parts.

The effect of a seed dressing lasts only about 5-7 weeks after sowing. Therefore, a foliar application is needed to protect the plant in the later stages of development especially as the tillers bear ears. A fungicidally active substance used successfully as a foliar treatment is triadimefon (Figure 1), another compound of the azole group (1, 3, 4). To study the uptake and transport of [benzene ring-U-^{14}C]triadimefon in barley leaves, 6 - 14 μg AI/leaf (equal to 225 g AI/ha) were applied in strips to the second developing leaf (29). Application was made to the basal part of the upper leaf surface when the leaf was almost completely developed. The untreated upper leaf parts were shielded with a plastic screen to prevent uptake of radiolabelled material via the gas phase.

Macroautoradiographs (Figure 5) of the leaves were taken at various times after triadimefon application and clearly showed translocation of ^{14}C-labelled compounds into the untreated leaf tip regions. Under field conditions, 28% of the radioactivity was found in the leaf tips within 12 days. However, in the greenhouse where environmental conditions (temperature, moisture and dew development) were more constant, 51% of the active substance translocated to the leaf tips as later confirmed by Buchenauer and Roehner (28). The difference between field and greenhouse results demonstrates the influence of precipitation and dew formation on the persistence of the active substance on the leaf as well as its uptake and internal transport. The treated leaf section as well as the untreated upper leaf part remained completely protected against infestation with mildew (Erysiphe graminis var. hordei). At the same time, protection against mildew was observed in the lower leaf region below the treated leaf strip although retranslocation into this part amounted to a maximum

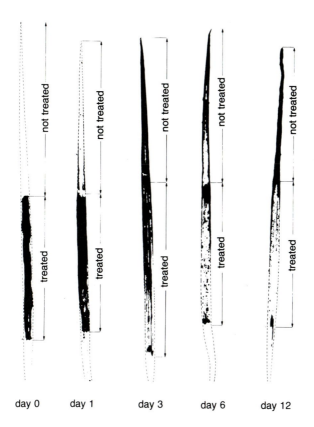

day 0   day 1   day 3   day 6   day 12

Figure 5. Translocation of radioactivity in spring barley after treating the upper side of the lower half of the leaf with [¹⁴C]triadimefon (macroautoradiographic radioactivity detection). (Reproduced with permission from Ref. 19. Copyright 1978 Pflanzenschutz-Nachrichten Bayer.)

of only 0.3% of the applied radioactivity. However, the protective effect or reduction in infestation was considerably weaker in these untreated basal regions.

Microautoradiographs of the barley leaf immediately after spraying showed the radioactive substance on the epidermis (Figure 6). However, a sequential time series of microautoradiographs of cryostat leaf sections (8 μm) shows that triadimefon and/or its metabolites are rapidly taken up (Figure 7). Subsequent transport takes place almost exclusively with the transpiration flow in the xylem. This transport system is then rapidly abandoned again so the active substance and/or metabolites move out of the transportation stream and are found in all leaf cells (29).

If the triadimefon spraying reaches the upper surface of the basal third of the leaf including the leaf sheath (Figure 8), then the transverse transport through the leaf sheath into the adjacent newly developed leaves is observed. In this way, up to approximately 20% of the active substance and/or its metabolites can migrate into the newly developing untreated leaves. The relatively mobile active substance or its metabolites then are rapidly transported within the leaf via the xylem, so that both the treated as well as the untreated parts of the leaf can become depleted with an accumulation of the active substance in the leaf tips and margins (Figures 5 and 8)(29).

Residue Situation in the Cereal Grain

This one-directional transport behavior results in a leaf treated with fungicide being protected as a whole. Active substance is transported into the apical region of newly developing leaves; however, redistribution into these leaves is limited because the active substance must first be transported downward in the phloem of the treated developed leaf before it can be transported via the xylem to the apical region of the newly developing leaf. For fungicidal compounds like triadimenol and triadimefon, which are predominantly if not exclusively transported in the xylem, this type of transport

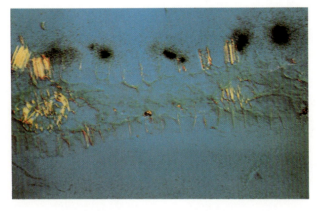

Figure 6. Microautoradiographic detection of [^{14}C]triadimefon immediately after being sprayed onto the epidermal cells of a barley (leaf cross section using interference phase contrast).

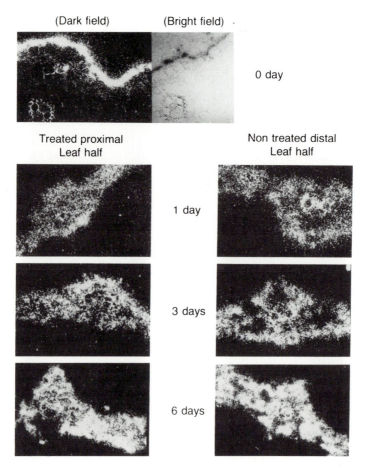

Figure 7. Radioactivity in barley leaf tissue after treating the upper side of the lower half of the leaf with [^{14}C]triadimefon (microautoradiograph, dark-field micrograph). (Reproduced with permission from Ref. 19. Copyright 1978 Pflanzenschutz-Nachrichten Bayer.)

behavior can be an advantage since the active substance and/or metabolites do not accumulate in the developing seeds or grain.

This is illustrated by data presented in Figure 9 (30). The
^{14}C- labelled fluotrimazole (Figure 1) was sprayed onto spring barley
in two lysimeters (0.96 m^2 each) using a chromatographic sprayer (30,
31). Spring barley at the K-stage prior to ear formation was sprayed
with either the agricultural use rate of 125 g/ha or a 2x rate (250
g/ha). At harvest, 57 days after treatment, about 1/3 of the applied
radioactivity was recovered in the straw, chaff and grain (Figure 9).
Active substance equivalent to 5.3 mg/kg of straw resulted from the
single dose and about twice as much active substance accumulated in
the residue treated with the 2x concentration. About 97% of the
total radioactive substances (fluotrimazole and metabolites) were
located in the leaves of the ripe spring barley with little radioactivity found in the stalks, chaff and grain.

The nature and distribution of the radiolabelled active substances and metabolites were determined using thin layer and gas
cochromatography in combination with liquid scintillation counting.
About 76% of the radioactivity found in the straw was fluotrimazole
and 14% was carbinol, the major metabolite (31). The awns, chaff and
grain, on the other hand, contained only traces of radioactivity,
less than 0.1% of the applied radioactivity (Figure 9) with the fluotrimazole equivalents in the grain computed to be less than 0.01
mg/kg for both spray concentrations. Fluotrimazole displays only a
locosystemic action and is probably only translocated in the xylem to
a very limited extent. The new growth, for example the ears, only
receives a small amount of active substance via the phloem.

Conclusions

These examples clearly show that it is not always desirable to develop fully systemic plant protection compounds, i.e. substances
which migrate equally well both in the xylem and in the phloem.
Transport in the xylem is often sufficient to effectively protect the
plant. Due to the physiology of the plant, the storage organs which

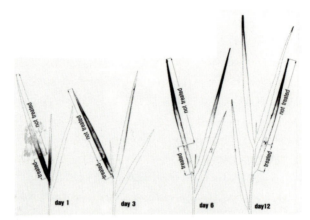

Figure 8. Uptake and translocation of radioactivity in spring barley after treating the upper side of the lower third of the leaf, including the leaf sheath, with [^{14}C]triadimefon (macroautoradiographic radioactivity detection). (Reproduced with permission from Ref. 19. Copyright 1978 Pflanzenschutz-Nachrichten Bayer.)

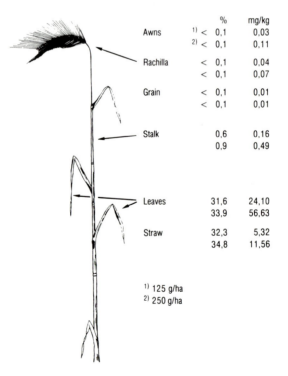

		%	mg/kg
Awns	1) < 0,1		0,03
	2) < 0,1		0,11
Rachilla	< 0,1		0,04
	< 0,1		0,07
Grain	< 0,1		0,01
	< 0,1		0,01
Stalk		0,6	0,16
		0,9	0,49
Leaves		31,6	24,10
		33,9	56,63
Straw		32,3	5,32
		34,8	11,56

1) 125 g/ha
2) 250 g/ha

Figure 9. Distribution of radioactivity and computed active substance equivalents in ripe spring barley 57 days after being treated with [^{14}C]fluotrimazol, sprayed radioactivity equals 100. (Reproduced with permission from Ref. 30. Copyright 1981 Verlag Eugen Ulmer.)

are used as fodder or for human nutrition are least contaminated. The translocation and accumulation of the fungicide could only be determined using the radiocarbon ^{14}C active substance.

The fluotrimazole results indicate that this is a very sensitive method. With the aid of radioactive labelling, it was computed that just 12.5 mg of active substance equivalents, i.e. compounds containing radiocarbon from fluotrimazole labelling, would be contained in the per hectare yield of 6,000 kg of spring barley (30, 31). Radioisotope labelling increases detection sensitivity to such an extent that even the slightest traces of residual carbon from an organic molecule can be quantified in the newly developing storage tissue. The active substance, with a correspondingly high specific radioactivity even when applied at a very low quantity, can still be characterized and identified by the combined methods of gas chromatography with mass spectrometry. For example, in an experiment where $[3-^{14}C]$ metamitron, a triazine herbicide, had been sprayed preemergence to sugar beets (32), about 25% of the radiocarbon found in the sugar beets at harvest (188 days after spraying) had been utilized to form saccharose, probably derived from mineralized $^{14}CO_2$ (33).

The recent studies with ^{14}C-labelled triazole fungicides attempt to gather information on residue analysis and biotests by applying macro- and microautoradiographic methods. Information about fungicide uptake and transport behavior in plants, the orders of magnitude of the active fractions and indications concerning the residue situation in the plant can be evaluated. The use of specialized radioisotope techniques in applied practical agriculture has been demonstrated. These studies not only improve the application of these compounds but also assist the consumer by learning more about the residue in his food after treatment with a chemical plant protectant.

Literature Cited

1. Büchel, K.H., Meiser, W., Krämer, W., Grewe, F., 8th International Congress of Plant Protection, Moscow, 1975, Section III, 111-118 (1975).

2. Grewe, F., Büchel, K.H., Mitt. Biol. Bundesanst. Land-Forst-wirtsch. (Berlin-Dahlem), 59, 652-655 (1975).

3. Kaspers, H., Grewe, F., Brandes, W., Scheinpflug, H., Büchel, K.H., 8th International Congress of Plant Protection, Moscow, 1975, Section III, 398-401 (1975).

4. Büchel, K.H., Krämer, W., Meiser, W., Brandes, W., Frohberger, P.E., Kaspers, H., 9th International Congress of Plant Protection, Washington, DC, 1979, Abstract of Paper No. 475 (1979).

5. Frohberger, P.E., Pflanzenschutz-Nachr. Bayer (Ger. Ed.) 31, 11-24 (1978).

6. Brandes, W., Kaspers, H., Krämer, W., Pflanzenschutz-Nachr. Bayer (Ger. Ed.) 32, 1-16 (1979).

7. Büchel, K.H., Plempel, M., In: J.S. Bindra, D. Lednicer (Eds.); Chronicles of Drug Discovery, Vol. 2, 235-269, John Wiley & Sons Inc. (1983).

8. Kuck, K.H., Scheinpflug, H., Tiburzy, R., Reisener, H.J., Pflanzenschutz-Nachr. Bayer (Ger. Ed.) 35, 209-228 (1982).

9. Smolka, S., Wolf, G. Pflanzenschutz-Nachr. Bayer (Ger. Ed.) 36, 97-128 (1983).

10. Kolbe, W., Pflanzenschutz-Nachr. Bayer (Ger. Ed.) 31, 38-59 (1978).

11. Förster, H., Buchenauer, H., Grossmann, F., Z. PflKrankh. Pfl-Schutz 87, 473-492 (1980).

12. Kolbe, W., Pflanzenschutz-Nachr. Bayer (Ger. Ed.) 35, 72-103 (1982).

13. Rawlinson, C.J., Muthyalu, G., Cayley, G.R., Plant Pathol. 31, 143-155 (1982).

14. Edgington, L.V., Peterson, C.A., In: M.R. Siegel, H.O. Sisler (Eds.): Antifungal Compounds, Vol. 2, 51-89, Marcel Dekker Inc., New York-Basel (1977).

15. Crowdy, S.H., In: R.W. Marsh (Ed.): Systemic Fungicides, 92-114, Longman, London-New York (1977).

16. Buchenauer, H., Untersuchungen zur Wirkungsweise und zum Verhalten verschiedener Fungizide in Pilzen und Kulturpflanzen. Habil-Schrift Univ. Bonn (1979).

17. Führ, F., In M. L'Annuziata, J.O. Legg (Eds.); Isotopes and Radiation in Agricultural Sciences Vol. 2, 239-270, Academic Press London (1984).

18. Führ, F., Rhein.-Westfael. Akademie der Wissenschaften, Vortraege N 326, 7-48, Westdeutscher Verlag Opladen (1984).

19. Steffens, W., Führ, F., Kraus, P., Scheinpflug, H., Pflanzenschutz-Nachr. Bayer (Ger. Ed.) 35, 171-188 (1982).

20. Führ, F., In: Agrochemicals: Fate in Food and the Environment, IAEA Vienna, 99-111 (1982).

21. Führ, F., Cheng, H.H., Mittelstaedt, W., Landw. Forsch. SH 32, 272-278 (1976).

22. Mückenhausen, E., Entstehung, Eigenschaften und Systematik der Böden der Bundesrepublik Deutschland, DLG-Verlag Frankfurt (1977).

23. Führ, F., Wieneke, J., Angew. Bot. 47, 97-106 (1973).

24. Wieneke, J., Führ, F., In: W.P. Duncan, A.B. Susan (Eds.), Synthesis and Applications of Isotopically Labelled Compounds. Proc. Int. Symp. Kansas City, MO, USA, 6-11 June 1982, Elsevier Scientific Publishing Co., Amsterdam, 373-374 (1983).

25. Biebl, R., Germ. H., Praktikum der Pflanzenanatomie, Springer, Wien-New York (1967).

26. Thielert, W., Steffens, W., Führ, F., Kuck, K.H. Pflanzenschutz-Nachr. Bayer (Ger. Ed.), 1985 (in preparation).

27. Thielert, W., Steffens, W., Führ, F., Scheinpflug, H. Pflanzenschutz-Nachr. Bayer (Ger. Ed.), 1985 (in preparation).

28. Buchenauer, H., Roehner, E., Z. PflKrankh. PflSchutz 89, 385-398 (1982).

29. Führ, F., Paul, V., Steffens, W., Scheinpflug, H., Pflanzenschutz-Nachr. Bayer (Ger. Ed.) 31, 116-131 (1978).

30. Steffens, W., Wieneke, J., Z. PflKrankh. PflSchutz 88, 343-354 (1981).

31. Wieneke, J., Steffens, W., Z. PflKrankh. PflSchutz 88, 385-399 (1981).

32. Mittelstaedt, W., Führ, F., Landw. Forsch. SH 37, 666-676 1981.
33. Müller, L., Mittelstaedt, W., Pfitzner, J., Führ, F., Jarczyk, H.J., Pestic. Biochem. Physiol. 19, 254-261 (1983).

RECEIVED October 1, 1985

The Pathogenesis of Plant Diseases
The Effect of Modern Fungicides

Hans Scheinpflug

Bayer AG, Agrochemical Division, Research and Development, Biological Research, D5090 Leverkusen, Federal Republic of Germany

Most of the older conventional fungicides such as the copper compounds, the dithiocarbamates, and the halogenalkylmercapto-imids (e.g. Captan) can be classified as broad-spectrum protectant biocides. In general, these fungicides are not absorbed or translocated by the plant and inhibit spore germination of fungal pathogens by contact activity. In order to prevent infection, they must be present on the surface of the plant at or before appearance of the pathogen and are suited for prophylactic use only, which requires a uniform film on the plant surface. These characteristics limit the possibility of successful disease control, and in many cases exclude it.

In contrast to these technically old-fashioned chemicals, compounds have been developed which differ in their chemical structure and mode of action. In addition to inhibiting spore germination, they interfere with different processes during the host colonization. These chemicals are absorbed and often translocated internally throughout the host plant. Translocated chemicals can have an inner-therapeutic action, stopping infection during and after the incubation period; in other words, they may act curatively or eradicatively. Histological and cytological studies were conducted to investigate the influence of several of these fungicides on various stages of fungal plant pathogens.

Over the last few years, a number of papers have described the influence of the azole fungicides on fungal growth and the important points of attack in the host-pathogen-system. In this paper several of these results will be presented.

Fungicide Mediated Morphological and Ultrastructural Changes of Fungi

Most of the work on the mode of action of the azole fungicides has been conducted with Ustilago species. In early experiments by Buchenauer (2), it was evident that triadimefon caused the sporidia of Ustilago avenae to lose their capacity for normal cell division. In nutrient solutions containing the fungicides, the sporidia, which normally separate after division, remained bound in mycelial-like clusters.

After treatment of U. avenae with triadimefon, nuarimol or other
compounds with the same mode of action, several distinct morphological changes could be seen at the ultrastructural level. This mainly
consisted of thickening of the peripheral cell wall and of incomplete
formation of septa, resulting in a lack of cell division of a multi-
branched mycelium.

Further morphological and ultrastructural changes could be
determined using thin sectioning and freeze etching:
a)   Considerable vacuolization and accumulation of lipid bodies
b)   Degeneration of intracellular organelles and membranes,
      especially the endoplasmatic reticulum
c)   Deformation of the plasmalemma
d)   Irregular aggregation of intramembrane particles
e)   Regular, hexagonal clustering of intramembrane particles

Richmond (12) made similar observations with Botrytis allii.
Pring (11) studied Uromyces vicia fabae infecting leaves of broad
beans and Puccinia recondita infecting wheat. Results in these tests
conducted in the intact host-pathogen system correlated well with the
changes observed in nutrient solution.

Influence of Triadimefon, Triadimenol and Bitertanol on the Pathogenesis of Several Fungal Diseases

Investigations with several host-pathogen-systems have shown that the
azole fungicides can have three effects:
a)   A direct influence on the pathogen by interfering with its
      sterol synthesis system
b)   An indirect influence on the pathogen by the reaction of
      the host plant
c)   An epidemiological effect by reducing the inoculum poten-
      tial

Influence on the Pathogenesis of Erysiphe graminis on Barley and
Wheat. In addition to conidiospores, which are formed in response to
weather conditions in shorter or longer generation cycles, cleisto-
thecia also bear spores which are formed as a result of a sexual
process and develop before the ripening of the cereals. After a
protective treatment with triadimefon, in concentrations ranging from
0.001% - 0.025 % a.i., and after a seed treatment with 7.5 g - 27.5 g
ai/100 kg seed, it was found that spore germination, as well as
development of appressoria of E. graminis f. sp. hordei were not
affected (13).

Certain changes in appearance of the appressoria were noted;
however, the function of the appressoria was not influenced (7).
Treatment did not affect the formation of the infection peg. Additionally, there was no influence of the treatment on the formation of
cell wall appositions, termed papillae, which are produced by the
host plant in response to the fungus beneath the site of penetration.

The first notable difference between treated and untreated
occurred when the fungus forms primary haustoria. At the beginning
of the normal formation of haustoria, which takes place just one day
after inoculation, the extrahaustorial membrane shows a light thickening and distension. Later on haustoria which develop in untreated

plants are characterized by well formed "fingers" (Figure 1). The effect of triadimefon as a spray application and of triadimenol as a seed dressing were basically the same, inhibiting the typical formation of these structures or they underwent abnormal development (Figures 2, 3, 4) (13).

On the fourth and fifth day after inoculation of the treated plants, stainable deposits form on the haustoria. These consist mainly of polysaccharides and are probably formed by the host plant (Figure 4). The material is the same as that in the papillae formed by the plant in response to the penetration infection peg. This encapsulation can be completed 4 to 8 days after inoculation and induces the fungus to stop development and blocks the formation of secondary haustoria (Figure 5) (13).

It is of special interest that the type of encapsulation seen in barley varieties with high resistance against powdery mildew is the same as that seen in the azole fungicide treated plants. This suggests that the fungicide induces some kind of resistance reaction in the host plant.

Encapsulation of haustoria could also be seen after curative fungicide treatment. Primary haustoria usually were not encased but all haustoria produced during and after the treatment were distinctly encapsulated, causing a complete inhibition of fungal growth and blocking spore formation.

Triadimefon at 250 ppm ai was sprayed on young mildew pustules. After 24, 48 and 72 hours, spores from treated pustules were harvested and tested for their capacity to germinate and cause infection (7). It was found, that the number of germinating spores which caused infection decreased with increasing time (Table I). The change in appearance of the mildew pustules after treatment is shown in Figures 6 and 7 (7).

Table I. Percent Infection of Barley by Conidia from
Bayleton-Treated Powdery Mildew Pustules,
in Relation to Time of Action.

Time of Action (Hrs.)	Percent E. graminis f. sp. hordei Infection*	
	BAYLETON	Untreated**
24	56	94
48	27	91
72	4	88

*Means from 3 experiments using 10 plants per pot with replications.
**Percent of leaf area infected.

Influence on the Pathogenesis of Various Puccinia Species of Small Grain Cereals. Wheat stem rust (Puccinia graminis var. tritici) is an example of a macrocyclic rust fungus with alternate hosts. The production of several generations of uredospores during the annual growth period provides the basis for an endemic appearance of the brown rust (Puccinia recondita), as well as for the development of an epidemic of stripe rust (Puccinia striiformis). Kuck et al. (6)

Figure 1. Haustoria of powdery mildew in an untreated plant.
Note the presence of numerous well-developed "fingers". (Repro-
duced with permission from Ref. 13. Copyright 1983
Pflanzenschutz-Nachrichten Bayer.)

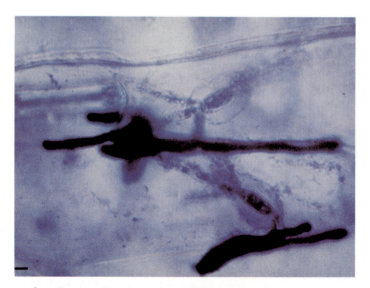

Figure 2. Haustoria of powdery mildew with abnormally formed
extrahaustorial membrane after protective application of 0.005%
Bayleton (3 days post-inoculation). (Reproduced with permission
from Ref. 13. Copyright 1983 Pflanzenschutz-Nachrichten Bayer.)

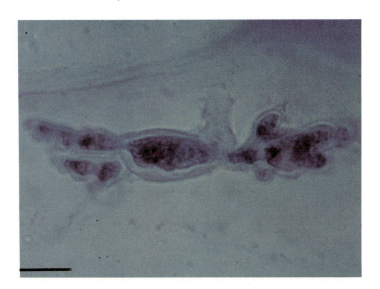

Figure 3. Deformed haustorium of powdery mildew with abnormally
formed "fingers" after treatment with 0.005% Bayleton (6 days
post-inoculation). (Reproduced with permission from Ref. 13.
Copyright 1983 Pflanzenschutz-Nachrichten Bayer.)

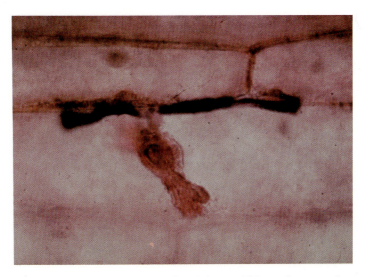

Figure 4. Encased haustorium of powdery mildew after seed treatment with Bayan 5 DS at 0.015 g/10 g seed (5 days post-
inoculation, combined staining with PAS and Coomassie Brilliant
Blue). (Reproduced with permission from Ref. 13. Copyright 1983
Pflanzenschutz-Nachrichten Bayer.)

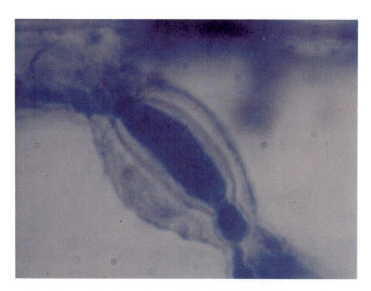

Figure 5.   Encapsulation of an 8-day-old haustorium of powdery
mildew after the protective treatment with Bayleton 25 WP
(stained with Coomassie Brilliant Blue). (Reproduced with permission from Ref. 13. Copyright 1983 Pflanzenschutz-Nachrichten
Bayer.)

Figure 6.   Pustule of E. graminis f. sp. hordei untreated (7 days
post-inoculation). (Reproduced with permission from Ref. 7.
Copyright 1979 Pflanzenschutz-Nachrichten Bayer.)

worked with stem rust (<u>Puccinia graminis</u> var. <u>tritici</u>) to study the influence of seed treatment with triadimenol and protective and curative leaf treatment with triadimefon on the disease development. Neither spore germination nor formation of appressoria were strongly affected by the different treatments. Even the correct location of the appressoria over a stoma was not disturbed (Figure 8).

During the typical infection process, a small tube emerges from the appressorium within 24 hours and, passing through the stomatal aperture, enters the substomatal cavity where the substomatal vesicle is formed. Intercellular hyphae growing from the vesicle produce mycelium which ramifies between host cells. Specialized terminal cells of the intercellular hyphae, the haustoria mother cells (HMC), produce a penetration peg that enters the host cell and gives rise to the haustorium.

One day after treatment the number of haustorial mother cells was not influenced in the triadimenol- and triadimefon-treated plants when compared with the untreated checks. Four days after inoculation, 75% of the untreated mycelia present in the host tissue formed more than five haustorial mother cells compared to 20-30% in treated tissue. In the time between 6 and 14 days after inoculation, the untreated mycelium had grown so intensely that it was impossible to count the HMC and sporulation had begun. Development of mycelia, which formed more than 5 haustorial mother cells, slowed down to a considerable extent between day 6 and day 14 in the treated plants.

With prolonged incubation of treated plants an increasing number of necrotic host cells could be visualized in epidermal and mesophyll cells. In only a few cases, the infection cycle proceeded to form sporogenic cells (6). Normally, haustoria do not stain bright blue using fluorescent dye; however, Kuck et al. (5) modified the procedure to allow observation of haustoria. Host cells autofluoresce a bright yellow when they become necrotic.

Figure 9 shows a germinating spore on a treated leaf surface, an appressorium positioned above a stoma and two epidermal cells which are attacked by the fungus. Their characteristic yellow stain indicates necrotic cells characteristic of a hypersensitive reaction.

The next figure (Figure 10) demonstrates that the intercellular mycelium in the treated tissue advanced to the mesophyll cells where it elicited a hypersensitive response within colonized, as well as in adjacent cells. The fungicide treatment induced in the susceptible wheat variety a reaction (hypersensitivity) which resembled that of a highly resistant host plant after rust attack, as described earlier for powdery mildew and barley.

Among the rust fungi, this observation was not only restricted to stem rust (<u>P</u>. <u>graminis</u> f. sp. <u>tritici</u>). Paul (9) could show a considerable difference between treated and untreated plants in the stage of substomatal vesicle formation of brown rust (<u>P</u>. <u>recondita</u>) and, more pronounced, during the development of intercellular mycelium and haustoria. Also, when infected wheat plants were treated with triadimefon prior to the opening of the rust pustules, these pustules failed to open. Further spreading of the fungus in the leaves was stopped (14).

<u>Influence of Bitertanol on the Pathogenesis of Venturia inaequalis</u>
<u>(Apple Scab)</u>. Apple scab is the most serious disease of apples and

Figure 7. Pustule of E. graminis f. sp. hordei treated with
Bayleton 4 days after and then photographed 7 days after inocula-
tion. (Reproduced with permission from Ref. 7. Copyright 1979
Pflanzenschutz-Nachrichten Bayer.)

Figure 8. Germinated rust spore that has developed an appressorium above a stoma. (Reproduced with permission from Ref. 6.
Copyright 1982 Pflanzenschutz-Nachrichten Bayer.)

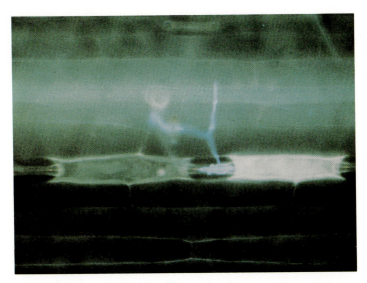

Figure 9. Three-day-old infection in wheat treated with Bayleton--germinating rust spore with an appressorium positioned above a stoma and two epidermal cells with hypersensitive reaction (fluorescence microscopy magnification X 190, photograph reduced 80%). (Reproduced with permission from Ref. 6. Copyright 1982 Pflanzenschutz-Nachrichten Bayer.)

Figure 10. Seven-day-old rust infection in wheat treated with Bayleton showing hypersensitive reaction of mesophyll cells (magnification X 190, photograph reduced 80%). (Reproduced with permission from Ref. 6. Copyright 1982 Pflanzenschutz-Nachrichten Bayer.)

is found in all areas of the world where apples are presently grown. In most cases, a well-timed chemical spray program has to be followed in order to control the disease. There is great interest in developing compounds with a curative effect, compounds which are able to control the disease after the establishment of the pathogen.

The disease cycle of apple scab (Venturia inaequalis) starts in spring with ascospores, which are discharged from asci, causing primary infection in the new leaves. During the summer, several generations of asexually formed conidia are produced which may cause infection in the same way as ascospores do.

Spore germination as well as appressorium formation were unaffected by bitertanol (Figure 11)(10). The dimension of the subcuticular stroma produced by the fungus was significantly smaller in bitertanol treated leaves. Three days after inoculation, numerous hyphae arose from the subcuticular stroma and kept growing between the cuticule and the outer cell wall of the untreated epidermal cells (Figure 12)(1).

The development of these hyphae was suppressed by the bitertanol treatment (Figure 13), and seven days after inoculation the stroma started to turn brown and to disintegrate (Figure 14). At the same time the fungus in the untreated leaves began to sporulate (1).

If leaves are treated with bitertanol by 72 to 96 hours after inoculation, the pathogen has already formed a stroma and the subcuticular mycelium has developed (Figure 15)(1). Three days after treatment one can see that the mycelium has stopped growing and the production of spores is suppressed (Figure 16)(10). Six days after the fungicide treatment, the stroma turns brown and starts to disintegrate (Figure 17)(1), but the fungus has already sporulated on the untreated leaves (Figures 18, 19)(10).

A fungicide treatment shortly before or at the beginning of conidia production caused a sharp reduction of conidia formation. This could be demonstrated by counting conidia of treated and untreated specimens. In addition to this, the treated lesions turned brown and became necrotic because the fungus had stopped growing (1).

Summary

Azole fungicides cause considerable morphological and ultrastructural changes of fungal plant pathogens. These effects occur not only under in vitro conditions, but they can also be observed in the intact host-pathogen-system. These fungicides act as protectants and as curative agents when applied against powdery mildew on cereals. Fungal spore germination and formation of appressoria remain almost unchanged after treatment. These fungicides cause encapsulation of haustoria. In this way, the uptake of nutrients by the fungus can be reduced or stopped.

Spore germination and appressoria formation during the pathogenesis of rust fungi is affected to a very low extent. The interference with haustorial formation is the most important effect. Colonized cells undergo a hypersensitivity reaction; finally they become necrotic. Further development of the obligate parasite is therefore stopped.

Spore germination of Venturia inaequalis, the apple scab fungus, is also unaffected by the studied azole fungicides. They prevent

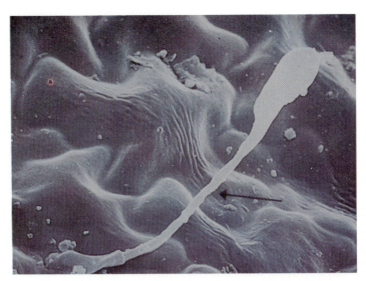

Figure 11. Germinating spore of <u>Venturia</u> <u>inaequalis</u> on a
bitertanol-treated leaf. (Reproduced with permission from Ref.
10. Copyright 1983 Pflanzenschutz-Nachrichten Bayer.)

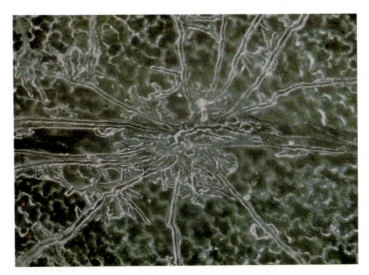

Figure 12. Development of the subcuticular stroma and mycelium
of <u>Venturia</u> <u>inaequalis</u> in an untreated leaf 3 days post-
inoculation. (Reproduced with permission from Ref. 1. Copyright
1981 Pflanzenschutz-Nachrichten Bayer.)

Figure 13. Suppressed development of the apple scab fungus in a leaf that was treated with bitertanol before inoculation (photographed 3 days post-inoculation). (Reproduced with permission from Ref. 1. Copyright 1981 Pflanzenschutz-Nachrichten Bayer.)

Figure 14. Stroma starting to turn brown and to disintegrate after treatment with bitertanol (7 days post-inoculation). (Reproduced with permission from Ref. 1. Copyright 1981 Pflanzenschutz-Nachrichten Bayer.)

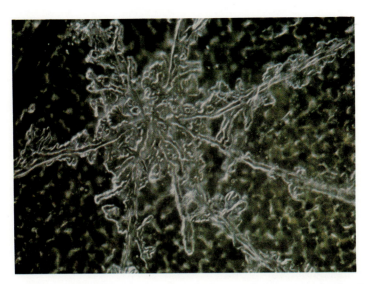

Figure 15. Stage of curative treatment with bitertanol showing subcuticular mycelium (3 days post-inoculation). (Reproduced with permission from Ref. 1. Copyright 1981 Pflanzenschutz-Nachrichten Bayer.)

Figure 16. Three days after the curative treatment with biter-tanol, the conidia are unable to pierce the cuticule (scanning electron microscopy, magnification X 600, photograph reduced 80%). (Reproduced with permission from Ref. 10. Copyright 1983 Pflanzenschutz-Nachrichten Bayer.)

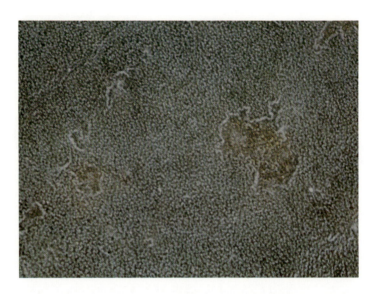

Figure 17. Six days after the bitertanol treatment, the stroma start to turn brown and disintegrate. (Reproduced with permission from Ref. 1. Copyright 1981 Pflanzenschutz-Nachrichten Bayer.)

Figure 18. Conidiophores and ripe conidia on the leaf surface of the untreated control 10 days post-inoculation (magnification X 600, photograph reduced 80%). (Reproduced with permission from Ref. 10. Copyright 1983 Pflanzenschutz-Nachrichten Bayer.)

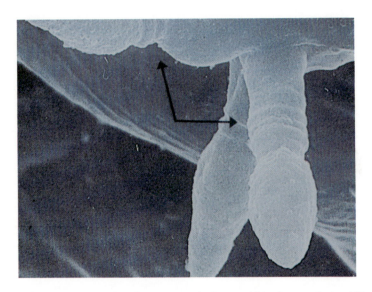

Figure 19. Ripe conidia borne on their conidiophores (magnification X 3300, photograph reduced 80%). (Reproduced with permission from Ref. 10. Copyright 1983 Pflanzenschutz-Nachrichten Bayer.)

subcuticular growth of the fungus as well as conida formation. Since this effect is also achieved during the incubation period, these fungicides are suitable for curative use. Because of their systemic properties and their eradicative effect, these products are ideal as chemical control agents and especially suitable for integrated pest management.

Literature Cited

1. Brandes, W., and Paul, P. (1981): Studies on the Effect of Baycor on Apple Scab Pathogenesis, Pflanzenschutz-Nachrichten Bayer 34, 48 - 59.
2. Buchenauer, H. (1976): Mechanism of Action of Bayleton (triadimefon) in Ustilago avenae, Pflanzenschutz-Nachrichten Bayer 29, 281 - 302.
3. Hippe, S. (1982): Ultrastrukturelle Verdnderungen in Sporidien von Ustilago avenae nach Behandlung mit systemischen Fungiziden, Dissertation Universität Hohenheim.
4. Hippe, S. (1983): Morphology of Ustilago avenae after Treatment with Systemic Fungicides as Studied by Scanning Microscopy, Phytopath. Z. 106, 321 - 328.
5. Kuck, H. H., Tiburzy, R., Hänssler, G., and Reisener, H. J. (1981): Visualization of Rust Haustoria in Wheat Leaves Using Fluorchromes, Physiological Plant Pathology 19, 439 - 441.
6. Kuck, K. H., Scheinpflug, H., Tiburzy, R., and Reisener, H. J. (1982): Fluorescence Microscopy Studies of the Effect of Bayleton and Baytan on Growth of Stem Rust in the Wheat Plant, Pflanzenschutz-Nacrichten Bayer 35, 209 - 228.
7. Paul, V., and Scheinpflug, H. (1979): Studies of the Effect of Bayleton on Barley Mildew, Pflanzenschutz-Nachrichten Bayer 32, 80 - 89.
8. Paul, V. (1981): Biology of Venturia inaequalis (Cooke) Winter, the Pathogen of Apple Scab, Pflanzenschutz-Nachrichten Bayer 34, 60 - 74.
9. Paul, V. (1982): Studies on the Effect of Bayleton on Pathogensis of Brown Rust Wheat (Puccinia recondita f. sp. tritici), Pflanzenschutz-Nachrichten Bayer 35, 229 - 246.
10. Paul, V., and Brandes, W. (1983): Scanning Electron Microscopy Studies on the Pathogensis of Apple Scab and its Control with Baycor, Pflanzenschutz-Nachrichten Bayer 36, 21 - 37.
11. Pring, R. J. (1984): Effects of Triadimefon on the Ultrastructure of Rust Fungi Infecting Leaves of Wheat and Broad Bean, Pesticide Biochemistry and Physiology 21, 127 - 137.
12. Richmond, D. V. (1984): Effects of Triadimefon on the Fine Structure of Germinating Conidia of Botrytis allii, Pesticide Biochemistry and Physiology 21, 74 - 83.
13. Smolka, S., and Wolf, G. (1983): Cytological Studies on the Mode of Action of Bayleton (triadimefon) and Baytan (triadimefon) on the Host-Parasite Complex - Erysiphe graminis f. sp. hordei, Pflanzenschutz-Nachrichten Bayer 36, 97 - 126.
14. Scheinpflug, H., Paul, V., and Kraus, P. (1978): Studies on the Mode of Action of Bayleton Against Cereal Diseases, Pflanzenschutz-Nachrichten Bayer 31, 101 - 115.

RECEIVED October 1, 1985

Progress in the Chemical Control of Diseases Caused by Oomycetes

F. J. Schwinn and P. A. Urech

Research and Development Department, Agricultural Division, Ciba-Geigy Limited, CH-4002 Basel, Switzerland

Among the fungi, the class Oomycetes is by number the smallest of the five classes, comprising some 70 genera with 500 species. However, in many respects, they are a unique and important class: 1) their cell walls differ from those of all other fungi, inasmuch as they contain cellulose instead of chitin; 2) they do not synthesize sterols; 3) their life cycle is diploid (in contrast to all other fungi); and 4) they form motile spores (zoospores).

Their classification is shown in Figure 1. From the phytopathologist's point of view, the Peronosporales are the most important order with the Peronosporaceae as the major family. Plant diseases caused by Peronosporales can be put into the three following groups: 1) foliar diseases, mainly downy mildews and late blight; 2) root and crown diseases, such as damping-off, seedling blights, root, collar and stem rot on annual and perennial crops; and 3) systemic diseases, that is diseases caused by infection of the roots from the soil or seed, distribution of the pathogen by the vascular system of the plant and manifestation of symptoms at the vegetation point or on the foliage. Most annual or perennial agronomical, horticultural and ornamental crops (such as grapes, potatoes, tobacco, tomato, hops, citrus, sunflowers, vegetables and soybeans), both in temperate and tropical climates, can be attacked by Peronosporales (Table I)(29, revised). The potential of the Peronosporales to result in epidemics within very short periods of favorable climatic conditions makes them an extremely devastating group of plant pathogens, the control of which has been a high priority for a long time and at present, amounts to about 25% of the total world fungicide market (Figure 2).

Until about ten years ago, protectant foliar fungicides (such as ethylene bis-dithiocarbamates and phthalimides) and soil sterilants (such as vapam or methylbromide) were the only chemical means of controlling diseases caused by Oomycetes. These compounds are nonspecific biocides affecting many vital cell processes of both the pathogen and the host plant. This means that they are non-selective,

0097–6156/86/0304–0089$06.00/0

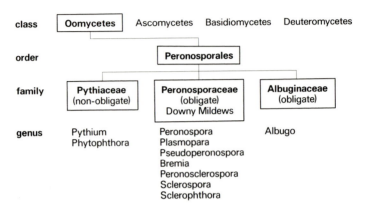

Figure 1. Plant pathogenic fungi.

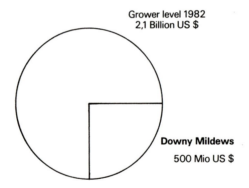

Figure 2. World fungicide market.

Table I.: Major Plant Pathogens in the Class Oomycetes (29)

Foliar Pathogens	Crop
Plasmopara viticola	grape vines
Phytophthora spp.	potatoes, tomatoes, cocoa
Peronospora tabacina	tobacco
Root and Crown Pathogens	Crop
Pythium spp.	sugar beets, vegetables, ornamentals
Phytophthora spp.	
	avocado, soybean, citrus, apples, ornamentals
Systemic Pathogens	Crop
Peronosclerospora/ Sclerospora/Sclerophthora	maize, sorghum, millet
Plasmopara halstedii	sunflower
Pseudoperonospora humuli	hops

and yet, can be used for the selective control of the pathogens. In the case of the soil sterilants, selectivity is achieved by the absence of the host plant at the time of application, that is, treatment before planting. The selectivity of the foliar protectants is based on their non-penetration into plant tissue. To be effective, they must be applied before infection occurs and must be present on the plant surface as long as it is susceptible. In view of their exposure to rainfall and weathering, this requires repeated applications at fairly high dosage rates. The limitations of these protective fungicides are obvious: they do not affect established local infections and cannot control systemic diseases. On the other hand, when used properly with narrow spray intervals and over the whole season, they have provided good protection from diseases over many decades and still continue to do so, as long as no serious epidemics develop.

Within the past ten years, the market introduction of several new types of fungicides has significantly improved the prospects of controlling the Oomycetes. They belong to five different chemical classes: the carbamates, the isoxazoles, the cyanoacetamide oximes, the etheyl phosphonates, and the acylalanines and related compounds. The chemical structures of those chemicals that have reached the commercial level are shown in Figures 3-5 (29, revised). Trade names, formulations and first reports are summarized in Table II (29, revised). The biological characteristics of these new fungicides and their impact on disease control have been reviewed by several authors (10, 16, 27, 28, 29, 33).

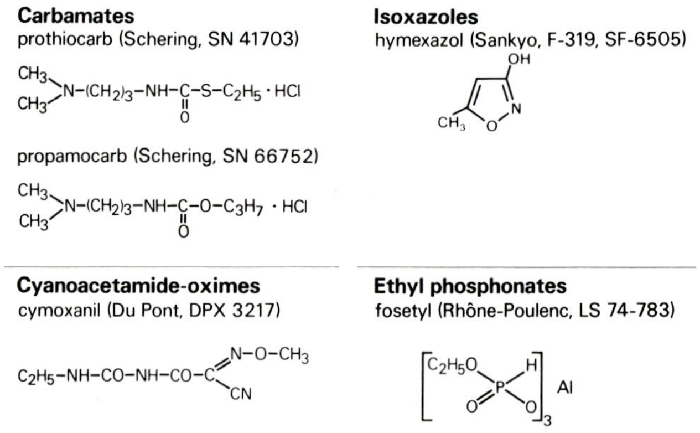

Carbamates
prothiocarb (Schering, SN 41703)

$$CH_3\diagdown N-(CH_2)_3-NH-\underset{\underset{O}{\|}}{C}-S-C_2H_5 \cdot HCl$$
$$CH_3\diagup$$

propamocarb (Schering, SN 66752)

$$CH_3\diagdown N-(CH_2)_3-NH-\underset{\underset{O}{\|}}{C}-O-C_3H_7 \cdot HCl$$
$$CH_3\diagup$$

Isoxazoles
hymexazol (Sankyo, F-319, SF-6505)

Cyanoacetamide-oximes
cymoxanil (Du Pont, DPX 3217)

$$C_2H_5-NH-CO-NH-CO-C\diagdown_{CN}^{N-O-CH_3}$$

Ethyl phosphonates
fosetyl (Rhône-Poulenc, LS 74-783)

Figure 3. Chemistry of new fungicides against Oomycetes (I). (29)

Phenylamides

Acylalanines
metalaxyl (Ciba-Geigy, CGA 48988)   furalaxyl (Ciba-Geigy, CGA 38140)

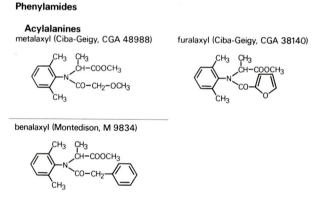

benalaxyl (Montedison, M 9834)

Figure 4. Chemistry of new fungicides against Oomycetes (II). (29)

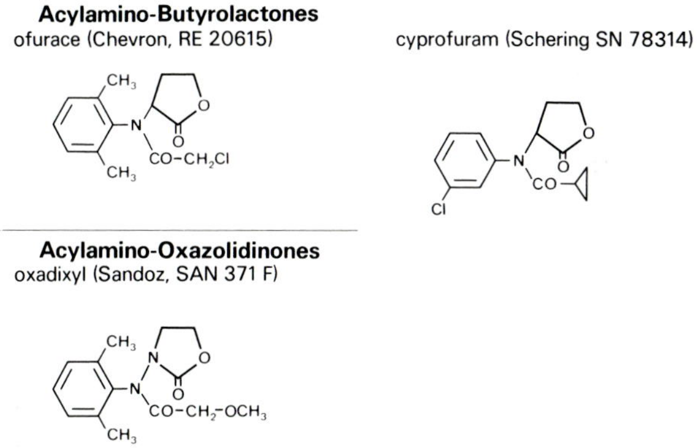

Acylamino-Butyrolactones
ofurace (Chevron, RE 20615)   cyprofuram (Schering SN 78314)

Acylamino-Oxazolidinones
oxadixyl (Sandoz, SAN 371 F)

Figure 5.   Chemistry of new fungicides against Oomycetes (III).(29)

Table II: New Chemicals for Control of Oomycetes (29)
(in order of their introduction)

Common Name	Trade Name(s) Formulation(s)	Chemical Group	First Report
prothiocarb	PREVICUR S 70 ; SCW	carbamates	1
hymexazol	TACHIGAREN	isoxazoles	42, 43
cymoxanil	CURZATE ; WP	cyanoacetamide-oximes	34
furalaxyl	FONGARID ; WP, G	acylalanines	30, 31
fosetyl Al	ALIETTE ; WP	ethyl phospho-nates	4
			46
metalaxyl	RIDOMIL ; WP, G ACYLON ; WP APRON ; SD	acylalanines	44
propamocarb	PREVICUR N ; SCW	carbamates	25
milfuram	PATAFOL ; WP	butyrolactones	23
benalaxyl	GALBEN ; WP, G	acylalanines	3
cyprofuram	VINICUR ; WP	butyrolactones	2
oxadixyl	SANDOFAN	oxazolidinones	19

The spectrum of activity of these compounds is shown in Table III. It varies from hymexazol, with an extremely narrow spectrum covering only the genus Pythium, to the phenylamides and related compounds covering all Peronosporales. The spectrum of fosetyl comprises even pathogens outside this order, such as Guignardia bidwellii (black rot of grapes), Phomopsis viticola (dead-arm of grapes) and Pseudopeziza tracheiphila (red fire of grapes). The reasons for these surprising differences are unknown.

Table III: Spectrum of Activity Against Peronosporales

Pathogens (Genera)	Fungicide and Activity*				
	prothiocarb/ propamo- carb	hymex- azol	cymox- anil	fosetyl	metalaxyl and related compounds
Pythium	+	+	+	+	+
Phytophthora					
on roots/stems	+	-	-	+	+
on foliage	-	-	+	-	+
Peronospora					
on roots/stems	+	-	-	-	+
on foliage	-	-	-	-	+
Pseudoperonospora	+	-	+	+	+
Plasmopara	-	-	+	+	+
Bremia	+	-	+	+	+
Peronosclerospora/ Sclerospora/ Sclerophthora	-	-	-	-	+
Albugo	-	-	-	-	+

* + = highly active
   - = no useful activity

Similiarly variable is the systemic activity of these compounds, that is the translocation in the vascular system of the plant. In this respect, cymoxanil with its very localized distribution, and fosetyl with its fast and strong translocation both acropetally and basipetally, are the extremes (Table IV). From this point of view, fosetyl is the most remarkable structure; it is the only commercial pesticide showing effective acropetal and basipetal translocation at normal use rates.

Table IV: Systemicity of the New Oomycetes Fungicides

| Chemical | Characteristics of Translocation | | |
	local (penetration)	acropetal	basipetal
prothiocarb/ propamocarb	++	+	-
hymexazol	++	+	-
cymoxanil	++	-	-
fosetyl	++	++	++
metalaxyl (and related compounds)	++	++	(+)

++ = strong, fast transport
+   = weak, slow transport
(+) = depending on crop species
-   = no transport in effective quantities

The new compounds are used at substantially lower rates than the conventional protectant fungicides (Table V), thus reducing the amount of chemicals brought into the environment. In addition, the acylalanines can be used as seed dressings at extremely low rates.

Table V: Comparison of Rates

Type of Disease	Rates of Standards	Rates of Systemics
Foliar	1.5 - 2.4 kg/ha (Dithiocarbamates)	ac: 0.25 kg/ha or 25 g/hl cy: 0.1 kg/ha or 12 g/hl fo: 1.5 kg/ha or 150 g/hl
Root and Crown	$2 - 4_2 g/m^2$ (Terrazole) $5 g/m^2$ (Captafol)	soil fo: $4 - 8 g/m^2_2$ soil ac: $0.2 - 2 g/m^2$ foliage fo: 300 g/hl seed ac: 17 - 35 g/100 kg
Systemic	-	seed ac: 70 - 210 g/100 kg

ac = acylalanines
cy = cymoxanil
fo = fosetyl

Despite the fact that the new compounds have been known and available for years, relatively little is known about their mode of action. Table VI reflects the state of the art. The lack of information is particularly surprising in the case of fosetyl and metalaxyl as far as their indirect effect is concerned, that is, the stimulation of the host plant's defense reactions (e.g. formation of phytoalexins). It is for the first time in the history of fungicides that such effects have been reported.

Table VI: Mode of Action of Oomycetes Fungicides

Compound	Mode of Action Physiological	Biochemical Level
prothiocarb/ propamocarb	Interference with membrane structure or function	Unknown
hymexazol	Delayed growth inhibition	Metabolite inhibiting RNA synthesis?
cymoxamil	Unknown	Inhibition of RNA synthesis
fosetyl	Direct fungitoxicity; secondary effects via plant defense	Unknown
metalaxyl	Direct fungitoxicity on mycelium; secondary effects via plant defense	Single site inhibition of RNA synthesis

The indirect mode of action has been postulated as the primary event in the case of fosetyl (5, 6, 11, 20, 26). However, recently Fenn and Coffey (17) showed that fosetyl and phosphoric acid (H_3PO_3), the degradation product of fosetyl in the plant, exhibit a strong fungitoxic effect in vitro if tested in a medium with low phosphate content. In addition, H_3PO_3 has a strong fungicidal effect in the plant. According to M. D. Coffey (1984, pers. comm.), fosetyl-resistant strains selected under laboratory conditions did not trigger the secondary effects described above. This would suggest that, in contrast to previous claims, the sequence of events in the host plant is the following: 1) a primary direct toxic effect on the pathogen, leading to a retardation or even cessation of fungal growth (fungistasis); 2) under such conditions, natural defense reactions of the host plant, normally too weak in susceptible cultivars, become effective and kill the pathogen; 3) thus, the death of the pathogen is due to combined action of the fungicide and the host plant.

When metalaxyl was used, necrotic reactions in susceptible crop cultivars similar to those of resistant ones, as well as phytoalexin accumulation, have been reported (12, 13, 21, 37, 41, 45). Here the use of metalaxyl-resistant strains could also clarify the significance of these findings, in particular concerning the accumulation of phytoalexins.

Spectrum of activity, crop tolerance and systemicity are the parameters defining the crop spectrum, the application method and the diseases controlled by the different types of compounds, as shown in Table VII.

Table VII: Main Uses of the New Oomycetes Fungicides

Compound	Main Target Crops	Main Usage Against	Application Methods
prothiocarb propamocarb	ornamentals, vegetables	diseases on roots and stems	drench, dip
furalaxyl	ornamentals		
hymexazol	rice, sugar beets		dust
cymoxamil	viticulture, agricultural crops, vegetables, (ornamentals)	foliar diseases	spray
fosetyl		foliar, root and stem diseases	spray, drench, dip, injection
metalaxyl and related compounds			spray, drench, dip, granule, seed dressing

From a practical and commercial point of view, the compounds shown in the lower part of Table VII are of much greater importance than those in the upper part. Whereas, prothiocarb, propamocarb, hymexazol, furalaxyl have to be regarded as specialty products for specialty crops, those listed in the lower part of the table have made broad applications in agriculture: cymoxanil due to its curative action which makes it a valuable mixing partner for protectants, fosetyl due to its upward and downward systemicity, and metalaxyl and the other phenylamides due to their high level of protective and curative activity, combined with a fast upward translocation in the plant. The overall progress in disease control achieved by them is shown on Table VIII.

Table VIII: Progress Achieved by the New Oomycetes Fungicides

Type of Disease	Comparative Performance	
	Standards	Systemics
Foliar	Good, if used protectively and intensively	outstanding (ac) to good (cy, fo): fast uptake/ distribution, curative action
Root and Crown	Unsatisfactory	Excellent after soil application (ac, fo) or foliar use (fo)
Systemic	Ineffective	Breakthrough (ac); high inherent activity, fast uptake by seedlings

ac = acylalanines
cy = cymoxanil
fo = fosetyl

Concerning the history and biological characteristics of the three product groups, my level of information is not uniform. This, and the fact that the acylalanines are the largest group, explains why more emphasis will be put on them in the remaining part of this paper. While there is little information about the history of the discovery of cymoxanil, there is evidence (4) that the fungicidal value of the ethyl phosphonates was discovered by incidental observation in a field trial. Thus, the initial step was a biological observation.

The history of the acylalanines began in herbicide research of Ciba-Geigy, Ltd., Basel, Switzerland. In the early seventies, a synthesis program was established aiming at new herbicides with a broader spectrum of biological activity in the class of anilides. The replacement of the alkoxyalkyl moiety of the rape seed herbicide dimethachlor (Teridox) by the ethyl propionate substituent, as found in the wild oats herbicide benzoylprop-ethyl (Suffix), led to a compound (CGA 22'574). This compound showed good herbicidal activity and, surprisingly, some systemic and curative fungicidal effect against late blight (Figure 6). This novel, yet weak, fungicidal performance of CGA 22'574 could have been easily overlooked if the corresponding methyl ester (CGA 29'212) had not been prepared in the course of the herbicide synthesis program. This molecule exhibited much stronger fungicidal action. It outperformed all standards in our fungicide screening against the Peronosporales, both in respect to residual and systemic behavior, as well as to protective and curative activity. However, its phytotoxicity was the limiting factor. In further synthesis programs, the herbicidal effect could be eliminated by replacing the chloroacetyl group by two other acyl fragments, namely the 2-furoyl group, thus leading to furalaxyl in 1973, and the methoxyacetyl group, leading to metalaxyl in 1974.

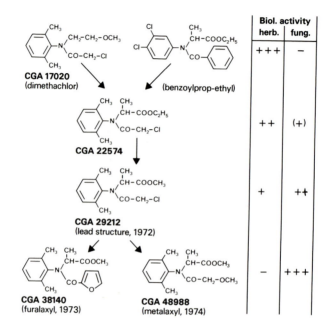

Figure 6.   History of acylalanines.

None of the many hundred analogues synthesized since then by Ciba-Geigy has led to a more powerful fungicide with as good a crop tolerance as metalaxyl (22). Since its introduction into the market in 1979, four chemically related compounds have been launched by other companies, as shown in Figures 4 and 5. Regarding their inherent fungicidal potency, it can be stated that none of them is more active than metalaxyl. Therefore in the following discussion, metalaxyl will be used as the best representative of the acylalanines.

Metalaxyl and most of its active analogues are chiral molecules. Chirality is caused by the asymmetric carbon atom in the alkyl side chain of the alanine moiety. The two optically pure enantiomers S (+) and R (-) differ widely in their biological activity both in vitro and in vivo. In all experiments, the R (-) enantiomer was more active than its antipode S (+) (22, 24, 30). The main characteristics of metalaxyl have been discussed in detail by several authors (7, 27, 28, 29, 32, 38). Of particular value is the rapid uptake of metalaxyl by the plant tissue, especially under the wet conditions that favor foliar Oomycete diseases. Acylalanines are easily translocated in the vascular system of the plant after foliar, stem or root treatment (35, 47). The predominant route of transport is the transpiration stream, thus apoplastic (12, 35). Symplastic transport occurs but is much less evident (35, 47). In potatoes treated by foliar sprays of metalaxyl concentrations (0.02-0.04 ppm), Bruin et al. (9) were able to demonstrate protection of harvested tubers from late blight.

Metalaxyl and related molecules are the only fungicides which so far control systemic diseases, such as downy mildew of hops (Pseudoperonospora humuli), black shank of tobacco (Phytophthora nicotianae var. nicotianae), downy mildew of sunflower (Plasmopara halstedii), and downy mildew of tropical maize, sorghum and millet (Peronosclerospora sorghi, Sclerospora graminicola). The acylalanines have a high inherent fungitoxicity. In vitro ED_{50} values for metalaxyl of typical target fungi are in the order of 0.01 to 1 ppm (7, 32). Based on circumstantial evidence, it can be assumed that the dose rates active on the cellular level in vivo are in a similar range.

The biological site of action of the acylalanines has been described by Staub et al. (37). In contrast to the classical protective fungicides which kill the germinating spores on the host surface, metalaxyl exhibits its fungitoxic effect only inside the host tissue. Neither zoospore or conidial germination, nor the penetration hypha or the formation of the initial haustoria is affected. In contrast, the further development of the pathogen is strongly and quickly inhibited. Regarding the development cycle of the pathogen, metalaxyl interferes with it over a much longer period of time than protective fungicides.

The biochemical mode of action has been studied by several authors (16, 18). It appears that metalaxyl inhibits RNA synthesis by interference with template-bound and α-amanitin-insensitive RNA polymerase action (15).

From the beginning of market introduction, the excellent performance of metalaxyl under variable climatic conditions and on a broad spectrum of crops made it very attractive to the farmer. It looked like a perfect solution for century-old problems had been found. However, since we live in an imperfect world, there is no such thing as a perfect solution. In the case of the acylalanines, resistance turned out to be the element of imperfection. It has developed in some areas, all outside the USA, in the following pathogens: Pseudoperonospora cubensis, Phytophthora infestans, Plasmopara viticola and Peronospora tabacina.

In view of the specificity and high effectiveness of the acyla-lanines, investigations to check the risk of resistance were initi-ated at early stages of its development. In model studies with Plasmopara viticola and milfuram, Lukens et al. (23) observed no change in sensitivity over nine successive generations. In training and spore mass selection experiments using Phytophthora infestans and metalaxyl, spontaneous mutants showing in vitro resistance were found. However, they had lost their pathogenicity. A large selec-tion trial through 14 disease cycles on potato plants in a greenhouse yielded no changes in sensitivity of the initial population (36). Therefore, it was concluded that there was no indication of a major risk of resistance. Additional studies on tobacco blue mold and on downy mildew of lettuce (Bremia lactucae) confirmed this conclusion.

In studies by Bruin (7), Bruin and Edgington (8) and Davidse (14) using chemical mutagens, strains of Phytophthora capsici and P. megasperma f. sp. medicaginis were found with both in vitro and in vivo activity and unchanged pathogenicity. Also, cross resistance to other acylalanines and related molecules, such as milfuram, was demonstrated (8).

At about the same time, the first information was received about product failure after continuous and exclusive use. The investiga-tion of the pathogen populations of such fields yielded highly resistant strains (28, 29, 40), whereas so far no field resistance has been reported for cymoxanil or fosetyl. Thus, in contrast to the favorable results of a broad range of model studies, resistance had appeared very fast under field conditions. The lesson to be learned from this experience is that results of model studies have to be used with caution. Model studies must include the use of chemical muta-gens and highly active, systemic fungicides should be used as if a risk of resistance exists until their mode of action is known.

After appearance of resistance in practice, metalaxyl as a single product for use against foliar diseases was withdrawn from the market. Strategies to reduce the risk of resistance were developed and introduced, based on prepack mixtures with fungicide partners with a different mode of action. Industry, under the auspices of GIFAP, formed a Fungicide Resistance Action Committee (FRAC), in order to coordinate the implementation of use strategies. This committee consists of working teams comprised of those companies having products of a particular chemical class, such as acylalanines, benzimidazoles, dicarboximides or ergosterol biosynthesis inhibitors.

With reference to the acylalanines, it can be stated that whenever they were introduced in mixtures, according to the recom-mended use concept, no cases of resistance leading to economic losses occurred. This use concept recommends a limited number of applica-tions at the time of serious damage potential combined with field monitoring programs (39). In the interest of the user, the manufacturer and the adviser, it is to be hoped that in the future all fungicides of this class will be used strictly on the basis of recommendations, which should help safeguard their availability. Concerning the other major new Oomycetes fungicides, cymoxanil and fosetyl, it can be stated that no resistance has yet been reported to date.

The practical uses of the new Oomycetes fungicides and the progress achieved by these products have been discussed previously (Table VII). It is worth mentioning that apart from propamocarb and prothiocarb, all compounds are used in combination with another fungicide: hymexazol in combination with metalaxyl, cymoxanil, or fosetyl, and the acylalanines in combination with protectants. In analyzing the reasons for these use concepts, the following can be said: 1) hymexazol with its extremely narrow spectrum needs a mixing partner to make it a useful broad spectrum rice fungicide; 2) cymoxanil with its purely curative action and its short persistence in the plant requires a protectant partner to make it a useful protectant product against the target pathogens; 3) fosetyl with its variability in performance against foliar diseases requires a partner to stabilize its effectiveness on a high level; 4) acylalanines need a partner to reduce the risk of resistance and to broaden the spectrum on crops like potatoes and grapes.

In conclusion, all the new systemic fungicides against Oomycetes had an immediate need for mixing partners. It can be assumed that this will also hold true for future systemic fungicides. This use concept calls for flexibility both on the part of industry, national registration authorities, and the extension services to make safe and powerful products available for plant protection.

Literature Cited

1. Bastiaansen, M.G., Pieroh, E.A., and Aelbers, E. 1974. Prothiocarb, a new fungicide to control Phytophthora fragariae in strawberries and Pythium ultimum in flower bulbs. Meded. Fac. Landbouwwet. Rijksuniv. Gent 39, 1019-1025.
2. Baumert, D. and Buschhaus, H. 1982. Cyprofuram, a new fungicide for the control of Phycomycetes. Meded. Fac. Landbouwwet. Rijksuniv. Gent 47, 979-83.
3. Bergamaschi, P., Borsari, T., Garavaglia, C. and Mirenna, L. 1981. Methyl N-phenyl-acetyl-N-2,6-xylyldl-alaûnate (M9834), a new systemic fungicide controlling downy mildew and other diseases caused by Peronosporales. Br. Crop Prot. Conf. 11th. 1, 11-18.
4. Bertrand, A., Ducret, J., Debourge, J.-C., and Horrière, D. 1977. Etude des propriétes d'une nouvelle famille de fongicides. Les monoéthylphosphites métalliques. Charactéristiques physicochimiques et propriétés biologiques. Phytiatr. Phytopharm. 26, 3-18.
5. Bompeix, G., Ravisé, A., Raynal, G., Fettouche, F., and Durand, M. 1980. Modalités de l'obtention de nécroses bloquantes sur feuilles détachées de tomate par l'action du tris-0-éthylphosphonate d'aluminium (phoséethyl d'aluminium), hypothèse sur son mode d'action in vivo. Ann. Phytopathol. 12, 337-351.

6. Bompeix, G., Fettouche, F., and Saurdressan, P. 1981. Mode d'action du phoséthyl Al. Phytiatr. Phytopharm. 30, 257-272.

7. Bruin, G.C.A. 1980. Resistance in Peronosporales to acylalanine-type fungicides. Ph.D. thesis Univ. Guelph, Ontario, Canada, 110 pp.

8. Bruin, G.C.A., and Edgington, L.V. 1980. Induced resistance to Ridomil of some oomycetes. (Abstr.) Phytopathology 70, 459-460.

9. Bruin, G.C.A., Edgington, L.V., and Ripley, B.D. 1982. Bioactivity of the fungicide metalaxyl in potato tubers after foliar sprays. Canad. J. Pl. Path. 4. 353-356.

10. Bruin, G.C.A., and Edgington, L.V. 1983. The chemical control of diseases caused by zoosporic fungi. In: Zoosporic plant pathogens. A modern perspective. (ed. S.T. Buczacki). Academic Press, London, 193-233.

11. Clairjeau, M., and Beyries, A. 1977. Etude comparée de l'action préventive et du pouvoir systémique de quelques fongicides nouveaux (phosphites-prothiocarb-pyroxychlore) sur poivron vis-à-vis de Phytophthora capsici (Léon). Phytiatr. Phytopharm. 26, 73-83.

12. Cohen, Y., Reuven, M., and Eyal, H. 1979. The systemic antifungal activity of Ridomil against Phytophthora infestans on tomato plants. Phytopathology 69, 645-649.

13. Crute, I.R. 1979. Lettuce mildew - Destroyer of quality. Agric. Res. Counc. (U.K.) Res. Rev. 5, 9-12.

14. Davidse, L. C. 1981. Resistance to acylalanine fungicides in Phytophthora megasperma f. sp. medicaginis. Neth. J. Pl. Path. 87, 11-24.

15. Davidse, L.C., Hofman, A.E., Velthuis, G.C.M. 1983. Specific interference of metalaxyl with endogenous RNA polymerase activity in isolated nuclei of Phytophthora megasperma f.sp.medicaginis. Exp. Mycol. 7, 344-361.

16. Davidse, L.C., and de Waard, M.A. 1984. Systemic fungicides. In: Advances in Plant Pathology (ed: D.S. Ingram and P.H. Williams), 2, 191-257.

17. Fenn, M.E., and Coffey, M.D. 1984. Studies on the in vitro and in vivo antifungal activity of fosetyl-Al and phosphorous acid. Phytopathology 74, 606-611.

18. Fisher, D.J., and Hayes, A.L. 1984. Studies of mechanism of metalaxyl fungitoxicity and resistance to metalaxyl. Crop. Prot. 3, 177-185.

19. Gisi, U., Harr, J., Sandmeier, R., Wiedmer, H. 1983. A new systemic oxazolidinone fungicide (SAN 371) against diseases caused by Peronosporales. Meded. Fac. Landbouwwet. Rijksuniv. Gent 48, 541-549.

20. Hai, V.T., Bompeix, G., and Ravisé, A. 1979. Rôle due tris-0-ethyl-phosphonate d'aluminium dans la stimulation des reactions de défense des tissus de tomate contre la Phytophthora capsici. C.R. Hebd. Seances Acd. Sci., Ser. D. 228, 1171-1174.

21. Hickey, E.L., and Coffey, M.D. 1980. The effects of Ridomil on Peronospora pisi parasitizing Pisum sativum: An ultrastructural investigation. Physiol. Plant Pathol. 17, 199-204.

22. Hubele, A., Kunz, W., Eckhardt, W., and Sturm, H. 1983. The fungicidal activity of acylanilines. In: IUPAC pesticide

chemistry. Human welfare and the environment. (ed. J. Miyamoto et al.), Pergamon Press, Oxford New York, 233-242.

23. Lukens, R.J., Cham, D.C.K., and Etter, G. 1978. Ortho 20615, a new systemic for the control of plant diseases caused by Oomycetes. Phytopath. News 12, 142.

24. Moser, H. and Vogel, C. 1978. Preparation and biological activity of the enantiomers of CGA 48'988, a new systemic fungicide. 4th Int. IUPAC Congr. Pestic. Chem., Abstracts II-310, Zurich.

25. Pieroh, E.A., Krass, W., and Hemmen, C. 1978. Propamocarb, ein neues Fungizid zur Abwehr von Oomyceten im Zierpflanzen- und Germüsebau. Meded Fac. Landbouwwet. Rijksuniv. Gent 43, 933-942.

26. Raynal, G., Ravisé, A., and Bompeix, G. 1980. Action du tris-0-éthylphosphonate d'aluminium (phoséthyl d'aluminium) sur la pathogénie de Plasmopara viticola et sur la stimulation des réactions de défense de la vigne. Ann. Phytopathol. 12, 163-175.

27. Schwinn, F.J. 1979. Control of Phycomycetes; a changing scene. Proc. Br. Crop Prot. Conf., 10th 3, 791-802.

28. Schwinn, F.J. 1981. Chemical control of downy mildews. In: The downy mildews (ed. D.M. Spencer). Academic Press, London New York.

29. Schwinn, F.J. 1983. New developments in chemical control of Phytophthora. In: Phytophthora: its biology, taxonomy, ecology and pathology (ed. D.C. Erwin, S. Barnicki-Garcia, P.H. Tsao). Amer. Phytopath. Soc., St. Paul, USA.

30. Schwinn, F.J., Staub, T., and Urech, P.A. 1977. A new type of fungicide against diseases caused by Oomycetes. Meded. Fac. Landbouwwet. Rijksuniv. Gent 42, 1181-1188.

31. Schwinn, F.J., Staub, T., and Urech, P.A. 1977. Die Bekämpfung Falscher Mehltaukrankheiten mit einem Wirkstoff aus der Gruppe der Acylalanine. Mitt. Biol. Bundesantalt Land Forstwirtsch. Berlin-Dahlem 178, 145-146.

32. Schwinn, F.J., and Staub, T. 1982. Biological properties of metalaxyl. In: Systemic fungicides and antifungal compounds (ed. H. Lyr and C. Polter). Akademie Verlag Berlin GDR, 123-133.

33. Schwinn, F.J., and Urech, P.A. 1981. New approaches for chemical disease control in fruit and hops. Proc. Brit. Crop Prot. Conf. Insecticides, fungicides 3, 819-833.

34. Serres, J.M., and Carraro, G.A. 1976. DPX-3217, a new fungicide for the control of grape downy mildew, potato blight and other Peronosporales. Meded. Fac. Landbouwwet. Rijksuniv. Gent 42, 645-650.

35. Staub, T., Dahmen, H. and Schwinn, F.J. 1978. Biological characterization of uptake and translocation of fungicidal acylalanines in grape and tomato plants. Z. Pflanzenkr. Pflanzenschutz 85, 162-168.

36. Staub, T., Dahmen, H., Urech, P., and Schwinn, F. 1979. Failure to select for in vivo resistance in Phytophthora infestans to acylalanine fungicides. Plant Dis. Rep. 63, 385-389.

37. Staub, T.H., Dahmen, H., and Schwinn, F.J. 1980. Effects of Ridomil on the development of Plasmopara viticola and Phytophthora infestans on their host plants. Z. Pflanzenkr, Pflanzenschutz 87, 83-91.

38. Staub, T., and Hubele, A., 1980. Recent advances in the chem-
     ical control of Oomycetes. Pages 389-422 in: Chemi der Pflan-
     zenschutz- und Schädlingsbekämpfungsmittel; Vol. 6. 4. Wegler,
     ed. Springer-Verlag, Heidelberg.
39. Staub, T., and Schwinn, F.J. 1980. Fungicidal properties of
     metalaxyl and its use strategies. Pages 154-157 in: Proc.
     Congr. Mediterr. Phytopathol. Union, 5th, Patras, Greece.
     Hellenic Phytopathol. Soc., Athens, Greece.
40. Staub, T., and Sozzi, D. 1983. Recent practical experiences with
     fungicide resistance. 10th Internat. Congr. Plant Path. Brigh-
     ton, U.K. 2, 591-398.
41. Staub, T.H., and Young, T.R. 1980. Fungitoxicity of metalaxyl
     against Phytophthora parasitica var. nicotianae. Phytopathology
     70, 797-801.
42. Takahi, Y., Nakanishi, T., Tomita, K. and Kaminura, S. 1974.
     Effects of 3-hydroxy isoxazoles as soil fungicides in relation
     to their chemical structure. Am. Phytopath. Soc. Japan 40.
     354-361.
43. Takahi, Y., Nakanishi, T., and Kaminura, S. 1974. Characteris-
     tics of hymexazol as a soil fungicide. Am. Phytopath. Soc.
     Japan 40. 362-367.
44. Urech, P.A., Schwinn, F., and Staub, T. 1977. CGA 48988, a novel
     fungicide for the control of late blight, downy mildews and
     related soil-borne diseases. Proc. Br. Crop Prot. Conf., 9th 2,
     623-6731.
45. Ward, E.W.B., Lazarovits, G., Stössel, P., Barrie, S.D., and
     Unwin, C.H. 1980. Glyceollin production associated with control
     of Phytophthora rot of soybeans by the systemic fungicide,
     metalaxyl. Phytopathology 70 738-740.
46. Williams, D.J., Beach, B.G.W., Horrière, D., and Maréchal, G.
     1977. LS 74-783, a new systemic fungicide with activity against
     Phycomycete diseases. Proc. Br. Crop Prot. Conf., 9th 2,
     565-573.
47. Zaki, A.I., Zentmyer, G.A., and LeBaron, H.M. 1981. Systemic
     translocation of C-labeled metalaxyl in tomato, avocado, and
     Persea indica. Phytopathology 71, 509-514.

RECEIVED October 1, 1985

Control of Fungal Plant Diseases by Nonfungicidal Compounds

Johan Dekker

Laboratory of Phytopathology, Agricultural University, 6709PD Wageningen,
The Netherlands

In recent years, there has been enhanced interest in more selective compounds for control of fungal diseases. Conventional fungicides, such as copper compounds and dithiocarbamates, are not very selective and are considered general plasma toxicants. Their use on the crop without appreciable harm to the plant, is mainly due to the fact that they remain on the plant surface. During the last two decades chemicals have been developed which are taken up by the plant and transported in its system, the so-called 'systemic' fungicides. Such compounds are considerably more selective since they inhibit fungal growth without being toxic to the plant. Another reason to search for more selective chemicals is the increasing concern about the hazards of toxic chemicals to man and the environment.

A further step towards selective action is the development of non-fungicidal compounds which interfere with the disease process without being directly toxic to the fungal pathogen or to other organisms. Such chemicals might act by increasing the resistance of the host plant or decreasing the virulence of the parasite. In addition to a highly selective action, these chemicals might also have other advantages, such as the induction of resistance in plants which lack specific genes for resistance, a longer lasting overall protection of the plant by immunization, and a reduced chance for the development of fungicide resistance.

During the search for systemic fungicides, chemicals are often found which provide disease control without a direct toxic effect on the fungus when grown on artificial medium. Moreover, it appears possible to introduce a pathogenic or non-pathogenic microorganism at a site and induce resistance at another site which indicates that chemical signals might be involved (16).

This paper will consider the basis for the principles of disease control by non-fungicidal compounds, the current research results, and the prospects for their future use. Induction of disease resistance by biotic agents will also be included because non-fungicidal substances might be involved in this process.

0097–6156/86/0304–0107$06.00/0
© 1986 American Chemical Society

Induction of Resistance

General. The defense of the host plants may be strengthened by various mechanisms. In some cases, natural disease resistance is based on the formation of a lignin barrier upon infection, which stops or retards the growth of the pathogen (13). Chemicals which stimulate the host to form such barriers during infection, increase the resistance of the host to disease. The formation of chemical barriers (phytoalexins) upon infection is quite common. Induced resistance may also be based on stimulation of a hypersensitivity response, strengthening of cell walls or membranes, insensitivity to fungal toxins or cell wall degrading enzymes, and/or the ability of the plant to inactivate the toxins and enzymes produced by the pathogen.

Elicitors of Phytoalexins. Each plant is naturally resistant to most pathogens which surround it. Therefore, it is logical to direct attention towards those substances which are thought to play a role in natural disease resistance. Many plants, upon contact with a pathogenic or non-pathogenic fungus, will form antifungal compounds called 'phytoalexins', this term meaning chemicals that protect the plant from parasites (15). The production of phytoalexins is induced by products of the pathogen, called elicitors. Application of elicitors to the plant to activate the plant's own defense mechanisms has been suggested. Although this sounds attractive, it might present problems. Phytoalexins are not innocuous chemicals. Excessive induction of these compounds in the plant may cause phytotoxicity and other undesirable side effects. Some of them are even potent toxicants to mammals. Moreover, it would be difficult and expensive to obtain elicitors, high molecular weight polysaccharides or glycoproteins, in sufficient quantity for practical application. Therefore, the possibilities for this approach seem very limited.

Various Synthetic and Natural Chemicals. There are an increasing number of reports about non-fungicidal compounds which provide disease control. The amino acids D- and DL-phenylalanine have been shown to reduce apple scab caused by Venturia inaequalis (17), and L-threo-β-phenylserine was active against cucumber scab caused by Cladosporium cucumerinum (1). The latter compound increased oxidative breakdown of indoleacetic acid in extract from cucumber seedlings, which suggests a relationship between chemotherapeutic effect and growth regulator metabolism. Amino acids not normally involved in the nitrogen metabolism of plants seem to be most active. An exception is L-methionine which protected cucumbers against powdery mildew (11). Its antifungal activity was antagonized by folic acid suggesting that it interferes with folic acid metabolism. Methionine and serine, which control bean rust, have also been shown to counteract the pathogen-stimulated incorporation of amino acids into protein (23). The mode of action of amino acids is still unknown.
Disease development often appears to be influenced by growth regulators. Most of these studies concern wilt diseases in various crops, especially caused by Fusarium spp. and Verticillium spp. (12). However, kinetin has been shown to be active against other types of diseases, such as the powdery mildews of cucumber (10) and tobacco (8). The mode of action of growth regulators in the disease control

is probably complex. Their value as plant protectants is doubtful because of phytotoxicity.

Indirect antifungal action has been reported for several unrelated compounds. Procaine hydrochloride, a local anaesthetic without fungitoxic activity, showed remarkable systemic action against various powdery mildews (9). Its mechanism of action is still elusive. It might effect the plant cell membranes, making the cell membrane more resistant to attack by fungal phospholipases (21). The protection of cucumbers against Cladosporium cucumerinum by the weak fungicide phenylthiourea might be due to an increase in polyphenols, favoring lignification at infection sites (14). Misato et al. (19) suggested an indirect action for soybean lecithin, a non-fungicidal substance with good activity against powdery mildew on strawberries and cucumbers.

According to Rathmell (24), a considerable number of compounds tested by the industry show indirect effects in disease control. However, in view of rather moderate disease control as compared to chemicals with a direct fungitoxic action, most did not receive additional testing. One exception is the work by Cartwright et al. (4) with the slightly antifungal compound 2,2 dichloro-3,3-methylcyclopropane carboxylic acid (WL 2835). Application of this compound to rice plants did not directly induce a demonstrable change in these plants. However, after inoculation with Pyricularia oryzae, two fungitoxic compounds appeared (momilactones A and B). These compounds are thought to play a role in increased resistance. The authors suggested that WL 2835 'sensitized' the host tissue so it reacted with a resistance response upon infection. Disease control by this and related compounds was not consistent enough to market them for commercial use.

Culture Filtrates. Schönbeck et al. (26) applied culture filtrates from various fungi and bacteria to beans and induced resistance to bean rust. The induction of resistance occurred at a distance from the site where the inducer was applied indicating it was systemic. A period of at least two days was required for the initiation of the resistance response. The phenomenon appeared to be nonspecific since the culture filtrates induced resistance in other plants and against various pathogens; however, only against obligate parasites. The active compounds in the culture filtrates have not been identified and their mechanism of action is unknown.

Biotic Agents. It is well-established that treatment of plants with virulent or avirulent forms of a pathogen, or with a non-pathogen, may induce the formation of fungitoxic compounds (phytoalexins) which prevent or retard subsequent infection by a pathogen. The phytoalexins are formed at the site of inoculation and are not transported to other plant parts.

In addition to this local resistance, biotic agents may induce systemic resistance. Infection of the first true leaf of cucumber plants with Colletotrichum lagenarium immunized other aerial plant parts against infection by C. lagenarium, Cladosporium cucumerinum and Pseudomonas lachrymans (16). Bean plants appeared systemically protected against Colletotrichum lindemuthianum when the first leaf had been inoculated with a non-pathogenic race of this fungus or with

C. lagenarium, which is not pathogenic to beans. Associated with the
immunization was an approximately threefold increase in peroxidase
activity, which occurred before inoculation with the challenge fungus
and at a distance from the site of inoculation with the inducer fungus. No phytoalexins could be detected in the protected plant parts
before the challenge inoculation, but after the challenge the phytoalexins rapidly accumulated at the infection sites (16). There is
evidence that immunization is the result of a signal transported from
the inducer leaf, but the nature of this signal is still unknown.
     The induction of resistance at a distance resembles the sytemic
elicitation of proteinase inhibitors in response to the infection of
tomato plants by Phytophthora infestans (22). Inoculation of the
lower leaves induced inhibition of proteinase activity in non-inocu-
lated upper leaves. The responsible agent is called the proteinase
inhibitor inducing factor (PIIF). Proteinase inhibitor activity increased about twofold after inoculation with an incompatible race
than compared with a compatible race. This proteinase inhibitor
appears associated with the hypersensitivity response and may inhibit
certain extracellular enzymes of the pathogen which are crucial for
virulence. The activity of PIIF appears to reside in oligosaccharides, which are enzymatically released from plant cell walls by the
pathogen (2). They might play hormone-like roles in regulating plant
defense responses at sites distant from their site of release (25).
De Wit and Bakker (30) also discovered a new protein in tomato leaves
at a distance from the site of inoculation with Cladosporium fulvum.
This protein also appeared associated with the hypersensitivity response.
     The study of factors or signals produced upon treatment of a
plant part with a biotic agent, and which cause metabolic changes
related to disease resistance at other sites, is of particular interest. Elucidation of their nature might provide new clues to manipu-
late the relationship between plant and parasite.

Commercial Compounds which Induce Resistance. Recently, a few
slightly or non-fungitoxic compounds, which are thought to control
disease primarily by the induction of resistance, have become available for practical use. Phosetyl-Al (aluminum tris ethylphosphonate)
was introduced in 1978 under the trade name Aliette. Although it is
only slightly fungitoxic in vitro, it provides remarkable control of
plant diseases caused by Oomycetes, particularly downy mildew and
Phytophthora diseases. Bompeix et al. (3) presented evidence that
phosetyl-Al stimulates the defense reactions of the plant by interference with polyphenol metabolism. In fungicide treated and subsequently inoculated plants, a necrotic zone appears which blocks or
retards colonization of the plant by the pathogen, allowing the plant
more time to mobilize other defense mechanisms. This hypothesis is
supported by the finding that induced resistance is counteracted by
certain enzymes which interfere with the processes leading to lignin
formation.
     Another promising compound is probenazole, 2-alloxy-1,2-benziso-
thiazole-1,1-dioxide, which was introduced under the name Oryzemate
for control of rice blast in Japan in 1977. Although only slightly
fungitoxic in vitro, it gave good disease control when applied to
paddy water. It caused a number of metabolic changes in the plant

which were thought to stimulate its defense reactions. It was pro-
posed that probenazole influenced the recognition between plant and
parasite which leads (via a chain of reactions) to the production of
chemical and physical barriers in the plant (27).

Reduction of Virulence

General. Some compounds, which exhibit no fungitoxic effects in
vitro, may interfere with the properties of the pathogen needed for
infection or for the build-up of an epidemic in the field. This may
involve pathogen enzymes used to degrade the plant cell wall and
cutin layer, pathogen toxins needed for colonization of host tissue,
or involve other products needed for infection and colonization.
Reduction of disease might also be caused by compounds which specifically block sporulation. Many compounds interfere with pathogen
virulence, but only a few do not inhibit fungal growth on artificial
medium and have sufficient results in vivo to be of practical use
(18).

Commercial Compounds which Reduce Virulence. The antibiotic validamycin A, produced by Streptomyces hygroscopicus var. limoneus, was
introduced in 1975 under the trade names Validacin and Valimon for
control of rice sheath blight caused by Pellicularia sasakii (29).
Under the trade name Solacol, it is available for control of Rhizoctonia solani on potatoes. It is not fungitoxic but alters the morphology of mycelium growth on artificial medium without reducing the
total mycelium mass (20). When treated plants are inoculated, the
pathogen appears unable to infect the plant tissue. Pathogenicity is
restored by the addition of myoinositol. The antibiotic may interfere with biosynthesis of myoinositol which appears indispensible for
pathogenicity of Rhizoctonia-type fungi.
    Another compound which reduces pathogenicity is tricyclazole,
5-methyl-1,2,4-triazolo[3,4-b]-benzothiazole, that was introduced in
1975 under the trade names Beam, Bim and Blascide for control of rice
blast. The primary site of action is between scytalone and vermelone
resulting in the inhibition of biosynthesis of melanins, which are
needed to make the appressaria of Pyricularia oryzae rigid enough to
penetrate the rice leaf. Rabcide, 4,5,6,7-tetrachlorophtahalide, introduced somewhat later for control of rice blast, also appears to
act at the same site in melanin biosynthesis as tricyclazole (5). A
similar mechanism of action is reported for several other rather
specific experimental compounds which control rice blast (31). Although tricyclazole inhibits melanization in Verticillium dahliae, it
does not control diseases caused by this pathogen (6). Rice leaves,
because of their silicium content, may require extra penetrating
power by the pathogen.

Development of Fungicide Resistance

Pathogens may become resistant to certain systemic fungicides and to
other fungicides which act at specific sites in the fungal metabolism. In practice, this has repeatedly led to the failure of disease
control by fungicides which originally were very effective.

Bompeix et al. (3) and Uesugi (28) have suggested that the use of
non-fungitoxic compounds, for example Aliette and Rabcide, might
result in less failure of disease control due to 'fungicide resistance'. However, in greenhouse experiments it was found that strains
of Phytophthora infestans and Pseudoperonospora cubensis resistant to
the systemic fungicide metalaxyl could not be controlled by application of Aliette (7). The basis for the cross-resistance between a
directly fungitoxic compound and a compound which presumably acts
indirectly by increasing the host plant resistance is not yet clear.
    The possibility of a pathogen overcoming chemically induced re-
sistance in the host plant is probably related to the mechanism of
action of the chemical inducer. If induced resistance in the host is
the result of the chemical acting at a single site in the host metabolism, it might be compared to vertical resistance, which is mono-
or oligogenic and can be overcome by the pathogen. However, if induced resistance involves a number of changes in the plant metabolism
it might be compared to polygenic horizontal resistance, which is
called durable resistance and is not easily overcome by the pathogen.
The same reasoning may hold for chemicals which decrease the virulence of the pathogen. In the latter case, the build-up of a resistant pathogen population may be counteracted by the competition of
sensitive strains which have not been eliminated by the non-fungitoxic compounds.

Future Prospects

Although compounds which do not inhibit fungi directly but control
disease have been found, only a few of them have been introduced into
practical use. Discovery of such compounds has been a matter of
trial and error and too little is known about their mechanisms of
action. More knowledge about the physiological-biochemical background of the host-parasite relationship will increase the chances
that compounds can be developed which interfere in a specific and
more efficient way with this relationship. These chemicals, depending on their mechanism of action, may have a lower chance for developing fungicide resistance that results in failure of disease control; however, generalization with respect to this phenomenon does
not seem justified.
    Another approach often suggested for disease control is the use
of compounds which are thought to play a role in natural disease
resistance. Examples of these compounds include phytoalexins, antifungal compounds formed upon contact of a plant with a fungus, and
elicitors, pathogen products which induce the formation of the phytoalexins. However, the overall induction of such compounds into the
plant may have phytotoxic and other undesirable side effects. The
production of these usually complex structures also appears to be
difficult and expensive. Finally, there are indications that chemicals may 'sensitize' the host tissue in such a way that a resistance
reaction results upon inoculation with a pathogen. Culture filtrates
of certain fungi and bacteria appear to contain substances which give
a similar effect. Studies on induced resistance following inoculation of certain microorganisms, have shown immunization at a distance
from the site of inoculation, indicating the involvement of signals.
The signals appear to move both upward and downward in plants, but

their nature is unknown. There is evidence that systemic induction
of resistance may also be obtained with chemicals.

The possibility exists that induced resistance in plants may be
longer lasting than the effect of fungicides, which may be broken
down or washed off by rain. Immunization of plant tissue, which
might take place at hormonal level, may require extremely low concentrations of a specific chemical. Increased research efforts are
needed to evaluate these speculations.

Literature Cited

1. Andel, O.M. van (1966) Amino Acids and Plant Diseases. Annual
   Review of Phytopathology 4, 349-368.
2. Bishop, P.D.; Makus, D.J.; Pearce, G.; Ryan, C.A (1981)
   Proteinase Inhibitor Inducing Factor Activity in Tomato Leaves
   Residues in Oligosaccharides Enzymatically Released From Cell
   Walls. Proceedings National Academy of Science USA 78, 3536-
   3540.
3. Bompeix, G.; Fettouche, F.; Saindreman, P. (1981) Mode d'action
   du phosétyl-Al. Phytiatrie-Phytopharmacie 30, 257-272.
4. Cartwright, D.; Langcake, P.; Ride, J.P. (1980) Phytoalexin Production in Rice and its Enhancement by a Dichlorocyclopropane
   Fungicide. Physiological Plant Pathology 17, 259-267.
5. Chida, T.; Uekita, T.; Satake, K.; Horano, K.; Aoki, K.; Noguchi,
   T. (1982). Effect of Fthalide on Infection Process of Pyricularia
   oryzae with Special Observation of Penetration Site of Appressoria. Annal Phytopathological Society of Japan 48, 58-63.
6. Chrysayi Tokousbalides, M.; Sisler, H.D. (1979) Site of Inhibition by Tricyclazole in the Melanin Biosynthetic Pathway of Verticillium dahliae. Pesticide Biochemistry and Physiology 11,
   64-73.
7. Cohen, Y.; Samoucha, Y. (1984) Cross-Resistance to Four Systemic
   Fungicides in Metalaxyl-Resistant Strains of Phytopthora infestans and Pseudoperonospora cubensis. Plant Disease 68, 137-139.
8. Cole, J.S.; Fernandez, D.L. (1970) Changes in the Resistance of
   Tobacco Leaf to Erysiphe cichoracearum D.C. Induced by Topping,
   Cytokinins and Antibiotics. Annals of Applied Biology 66, 239-
   243.
9. Dekker, J. (1961) Systemic Activity of Procaine Hydrochloride on
   Powdery Mildew. Netherlands Journal of Plant Pathology 67, 25-
   27.
10. Dekker, J. (1963) Effect of Kinetin on Powdery Mildew. Nature,
    London 197, 1027-1028.
11. Dekker, J. (1969) L-methionine Induced Inhibition of Powdery
    Mildew and Its Reversal by Folic Acid. Netherlands Journal of
    Plant Pathology 75, 182-185.
12. Erwin, D.C. (1977) Control of Vascular Pathogens. In: Antifungal
    Compounds (editors M.R. Siegel and H.D. Sisler), Marcel Dekker
    Inc., New York, Vol. 1, 163-224.
13. Hijwegenn, T. (1963) Lignification, A Possible Mechanism of Ac-
    tive Resistance Against Pathogens. Netherlands Journal of Plant
    Pathology 69, 314-317.

14. Kaars Sijpesteijn, A.; Sisler, H.D. (1968) Studies on the Mode of Action of Phenylthiourea, A Chemotherapeutic for Cucumber Scab. Netherlands Journal of Plant Pathology 74, Supplement 1, 121-126.
15. Kuč, J. (1977) Phytoalexins. In: Physiological Plant Pathology (editors R. Heitefuss and P.H. Williams), Springer-Verlag, Berlin, Heidelberg, New York, 632-652.
16. Kuč, J. (1983) Induced Systemic Resistance in Plants to Diseases Caused by Fungi and Bacteria. In: The Dynamics of Host Defense (editors J.A. Bailey and B.J. Deveerall), Academic Press, Australia. 191-221.
17. Kuč, J.; Barner, E.; Daftsios, A.; Williams, E.B. (1959) The Effect of Amino Acids on Susceptibility of Apple Varieties to Scab. Phytopathology 49, 313-315.
18. Langcake, P. (1981) Alternative Chemical Agents for Controlling Plant Disease. Philosophical Transactions Royal Society London B295, 83-101.
19. Misato, T.; Homma, Y.; Ko, K. (1977) The Development of a Natural Fungicide, Soybean Lecithin. Netherlands Journal of Plant Pathology 83, Supplement 1, 395-402.
20. Nioh, T.; Mizuschima, S. (1974) Effect of Validamycin on the Growth and Morphology of Pellicularia sasakii. Journal of General and Applied Microbiology 20, 373-383.
21. Papahadiopoulos, D. (1973) Studies on the Mechanism of Action of Local Anesthetics with Phospholipid Model Membranes. Biochimica Biophysica Acta 265, 196-186.
22. Peng, J.H.; Black, L.L. (1976) Increased Proteinase Inhibitor Activity in Response to Infection of Resistant Tomato Plants by Phytophthora infestans. Phytopathology 66, 958-963.
23. Pozsar, B.I.; Kristev, K.; Kiraly, Z. (1966) Rust Resistance Induced by Amino Acids: A Decrease of the Enhanced Protein Synthesis in Rust-Infected Bean Leaves. Acta Phytopathologica 1, 203-208.
24. Rathmell, W.G. (1980) Active Defense and Crop Protection. NATO Advanced Study Institute, Cape Sounion, Greece.
25. Ryan, C.A.; Bishop, P.; Pearce, G.; Darvill, A.G.; McNeil, M. Albersheim, P. (1981) A Sycamore Cell Wall Polysaccharide and Chemically Released Tomato Leaf Polysaccharide Possess Similar Proteinase Inhibiting-Inducing Activities. Plant Physiology 68, 616-618.
26. Schönbeck, F.; Dehne, H.W.; Beicht, W. (1980) Untersuchungen zur Aktivierung unspezifisher Resistenz mechanismen in Pflanzen. Zeitschrift fur Pflanzenkrankheiten und Pflanzenschutz 87, 654-666.
27. Sekizawa, Y. (1983) Mode of Action of Rice Blast Protectant Probenazole. In: Pesticide Chemistry, Human Welfare and the Environment, Vol. 3: Mode of Action, Metabolism and Toxicology (editors J. Miyamoto and P.C. Kearney), Pergamon Press, 147-152.
28. Uesugi, Y. (1982) Pyricularia oryzae of Rice. In: Fungicide Resistance in Crop Protection (editors J. Dekker and S.G. Georgopoulos), Pudoc, Wageningen, 207-218.
29. Wakae, O.; Matsuura, K. (1975) Characteristics of Validamycin as a Fungicide for Rhizoctonia Disease Control. Review of Plant Protection Research 8, 81-92.

30. Wit, P.J.G.M. de; Bakker, J. (1980) Differential Changes in Soluble Tomato Leaf Proteins After Inoculation With Virulent and Avirulent Races of Cladosporium fulvum (syn. Fulvia fulva). Physiological Plant Pathology 17, 121-130.
31. Woloshuk, C.P.; Sisler, H.D. (1982) Tricyclazole, Pyroquilon, Tetrachlorophthalide, PCBA, Coumarin and Related Compounds Inhibit Melanization and Epidermal Penetration by Pyricularia oryzae. Journal Pesticide Science 7, 161-166.

RECEIVED October 1, 1985

Control of Cereal Diseases with Modern Fungicides in Western Europe

G. M. Hoffmann

Faculty of Agriculture, Technical University Munich, D8050 Freising-Weihenstephan, Federal Republic of Germany

Economic and social changes in Western Europe in the past 35 years have caused farmers to alter farming practices in an attempt to achieve higher yields. These farmers have increased capital investments and decreased man-power hours needed for farming. This point is evidenced by a 50% reduction in the number of persons working in agricultural fields between 1970 and 1982 (Table I), as Western Europe entered the era of high intensity crop production(1).

Table I. Percentage of Population Employed in Agriculture.

Country	1970	1982
Denmark	11.1	6.4
France	13.7	7.9
Germany FR	7.5	3.5
Netherlands	8.1	4.9
United Kingdom	2.8	1.9

This change in farming practices can also be observed by comparing labor units per hectare of land used in agriculture. In the Federal Republic of Germany, labor units/ha were 29 in 1950 and only 7.9 in 1982. This reduction parallels an increase in combine harvested crops, especially winter cereals (wheat and barley).

In Germany, the cropping area for winter wheat increased from 1.1 million hectares in the mid-1950's to 1.6 million hectares in 1980. Winter barley production increased from 127,000 hectares to 1.3 million hectares in 1982. This shift to a small grain monoculture limited the possibilities for increasing crop rotation and increased the need for crop protection chemicals. Despite the added pest pressures from the lack of crop rotation, yields in Germany have increased by over 100% from 1952 to 1980 (Figure 1)(9). This increase in yield is primarily due to the selection and growth of cereal varieties with a genetic potential for high yields (140 dt/ha). Other reasons for yield increase include: earlier seeding,

0097–6156/86/0304–0117$06.00/0

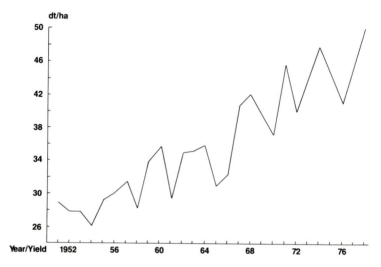

Figure 1. Yield increase in wheat in the Federal Republic of Germany from 1952 to 1980. (Reproduced with permission from Ref. 9. Copyright 1981 DLG Verlag Frankfurt.)

higher plant populations/ha, higher mineral nutrition, and the use of growth regulators to reduce lodging.

True yield potentials can only be expressed when the control of weeds, pathogenic agents, and insect pests are coupled with proper cultivation, fertility, and other plant protection measures. An example of these high yields in Western Europe appears in Table II (1).

Table II. Average Yield of Wheat and Barley (dt/ha)
in Western Europe 1982.

Country	Wheat	Barley
Denmark	67.07	42.90
France	52.32	41.93
Germany FR	54.71	46.81
Netherlands	73.90	56.72
United Kingdom	61.95	49.11
United States	23.96	30.84

This system of integrated crop production places high demands on all aspects of crop production, including the controlled use of pesticides. Development and benefit from integrated pest management systems depend upon the combined efforts of the researcher, development and scouting persons and a highly trained farmer.

Perhaps the most important restrictive negative factors in cereal production in Western Europe are the fungal diseases. Epiphytotics of several diseases are a common occurrence. Other diseases such as powdery mildew and glume blotch only occur in the more humid regions. Fungicides used for the control of foliar diseases on wheat (Table III) were primarily developed in the last decade. In the past 10 years, triazole and amine derivatives have played a significant role in disease control in cereals. These fungicides have gained wide-spread acceptance in only a few years because they are highly efficacious and ecologically safe.

The portion of cereal cropping areas which have been sprayed with fungicides within the last few years is hard to ascertain. Jenkins and Lescar (3) estimated the following percentages for 1979: Denmark-23%, France-27%, Netherlands-40%, Germany (F.R.)-29%, and the U.K.-50%. An international survey on the use of fungicides on cereals produced in 1983 is compiled in Table IV. These figures were confirmed by Priestly and Bayles (8) for the U.K.

In France and Germany, disease control measures on cereals are directed primarily against foot rot and leaf diseases, or leaf and ear diseases. In the United Kingdom, treatments are generally made for the control of leaf diseases, such as powdery mildew. It should be mentioned that these data do not include the use of the azole-compound triadimenol as a seed dressing for early season protection against powdery mildew and rust fungi on wheat and barley.

Table III. Some Serious Fungal Diseases on Cereals in Western
Europe and Fungicides Used for Control.

Crop	Disease	Fungicides
Wheat	Powdery mildew	Fenpropimorph, Triadimenol, Propiconazole
	Glume blotch	Ortho-Difolatan, Propiconazole, Prochloraz
	Eyespot	Carbendazim, Thiophanate-methyl, Prochloraz
Barley	Powdery mildew	Fenpropimorph, Triadimenol, Propiconazole, Prochloraz
	Net blotch	Propiconazole, Prochloraz
	Typhula blight	Bitertanol, Triadimenol

Table IV. Fungicide Treatment in Cereal Cropping Areas
(x 1000 ha) Accumulated for Diseases in
Western Europe 1983.

Country		Treatment Against			
	Foot Rot	Foot Rot and Leaf Diseases	Leaf Diseases	Leaf and Ear Diseases	Ear Diseases
Denmark	201	33	279	1,757	14
France	1,575	4,435	-	4,880	-
Netherlands	20	-	141	-	218
United Kingdom	1,180	485	2,400	565	1,420
Germany FR	534	1,392	1,098	1,387	249
Sum	3,510	6,345	4,197	8,589	1,901

Fungicide effectiveness is based on the severity of disease
pressure, the susceptibility of the host, the biological environment,
the biological activity of the compound and the time and number of
applications. In Western Europe, losses in cereal yields range from
20 to 30% (10-20 dt/ha). The average loss in many trials due to
cereal disease lies between 5 and 12%. However, these losses can be
reduced through the use of fungicides. Survey data from France (5)
demonstrate the yearly variations of the effects of fungicide treatments on winter wheat (Table V).

Table V.   Mean Yield Increases of Winter Wheat in France after
Use of Fungicides.

Year	Yield Increase (%)	Frequency of Yield Increase > 0.5 t/ha (%)
1973	7.3	37
1974	8.2	40
1975	6.5	29
1976	2.0	4
1977	9.2	48
1978	10.8	60
1979	6.4	31

The percent increase in yield is often related to weather
conditions. In dry years (1976) a relatively small number of fungicide cereal trials showed an increase in yield, but during wet years
(1978) 60% of the tests showed yield increases. In long-term trials
(4) in the Western region of Germany, seed treatments and foliar
applications of triazole fungicides reduced crop losses from powdery
mildew up to 25% (Table VI).

Table VI.   Long-term Results of Mildew Control
on Winter Wheat 1975/76 to 1980/81.

Site	Treatment	Grain Yield Absolute (dt/ha)	Relative (%)
A	No treatment	48	100
	Baytan special (seed treatment)	56	118
	Baytan special + Bayleton	60	125
B	No treatment	53	100
	Baytan special (seed treatment)	57.2	108
	Baytan special + Bayleton	62.0	117

Efficacy evaluations (7) of plant protectant compounds need to
be made in various cropping systems using varying levels of nitrogen
(Table VII). Controlling eyespot alone (PS_2) is inadequate. When
combined with treatments against leaf and ear diseases (PS_3, PS_4),
yield increases of up to 12% have been observed. The additional use
of organo-phosphorous insecticides (PS_5) against aphids results in
further increases. There is a 20.7% difference in yield between

cropping system IA and VC. In addition to the effect of fungicides
on yield, the quality of grain (specific weight = kg/hl) can be
improved.

Table VII. Different Types of N-fertilization, Disease Control
Measures and Yield (dt/ha) of Winter Wheat
(29 Trials 1980 - 1982)

Cropping System	Disease Control	Type of N-Nutrition					
		A		B		C	
		Abs.	Rel.	Abs.	Rel.	Abs.	Rel.
I	PS_1	69.3	100	71.0	100	71.5	100
II	PS_2	70.8	102.6	72.8	102.5	73.2	102.3
III	PS_3	72.6	105.2	74.4	104.7	75.0	104.8
IV	PS_4	77.7	112.6	70.3	11.6	79.5	111.1
V	PS_5	70.9	115.7	81.6	114.9	83.3	116.5

PS_1	=	Cycocel (Growth regulator)
PS_2	=	Cycocel + Cercobin Fl
PS_3	=	Cycocel + Cercobin Fl + Corbel
PS_4	=	Cycocel + Cercobin Fl + Corbel + Ortho-Difolatan
PS_5	=	Cycocel + Cercobin Fl + Corbel + Ortho-Difolatan + Parathion POX

There is no doubt that use of fungicides must be justified on an
ecological and economical basis. The farmer's decision to use a
fungicide is based on the results of many years of local experience
and the occurrence of specific disease organisms of economic importance in the growing area. On the other hand, it is the task of
plant pathologist and counseling services to compile basic knowledge
on epidemiology of certain pathogens and on the calculation of
disease/loss rate. These efforts are combined to provide warning
forecasts to the farmer.
    In England, it is recommended that fungicidal treatment for
control of powdery mildew on spring barley be applied when 3% of the
area of the two oldest green leaves is infested with mildew. For
winter wheat in the Federal Republic of Germany, the threshold value
is about 1% infested area of the uppermost two leaves or an infestation frequency of more than 60%, that is when about two-thirds of all
plants show mildew attack. These values are thought to indicate the
presumable beginning of an epidemic development.
    Models for risk estimation of cereal eyespot were developed
which employ a great number of parameters (previous crop, soil,
weather, cultivation, sowing date, plant density, cultivar and vigor
in spring). Risk models provide a good orientation for the counseling services and can be adjusted to local conditions and experiences.
An example of a well-developed prognostic model on cereal diseases is
the cooperative project EPIPRE (Figure 2). The heart of EPIPRE is a
data bank which contains the core data for each field and the variable data provided by the farmers during the season. EPIPRE can also
monitor recent weather data. The computation procedures for risk

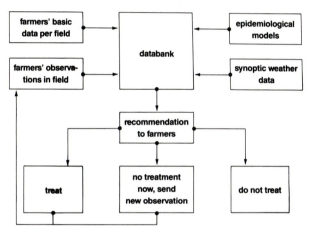

Figure 2. Prognostic model on cereal diseases. (Reproduced with permission from Ref. 10. Copyright 1981 Organization Europeenne et Mediterraneenne pour la Protection des Plantes.)

estimation have been stored in the model section. Linked to the epidemic growth model is a complex decision network with 27 decision points. The decision network uses all available epidemiological, agronomic and economic information. The number of fields registered under this program in the Netherlands was about 1200 in 1981 (10).

Although generally satisfactory for mildew and rust, disease thresholds have not been adequate for eyespot and diseases caused by Septoria, Drechslera and Rhynchosporium. Various prognostic models are under intensive investigation in different institutions. They have to be compiled for every disease and adapted to regional conditions. For this reason, no single European model can be expected, since ecological and economical pre-requisites between the various countries and within each country, as well as between the various regions, are too different. Numerous imponderabilities, especially the lack of accurate weather prognoses, frequently result in routine treatments by the farmers. These fungicide treatments have been shown to be economically justified by the average of many years, but not in every single year.

The use of modern fungicides has also to be viewed under the aspect of the energy output/input ratio. Agricultural and forest economies are the only economic activities which result in a net production. However, only 0.5% of the radiant energy reaching the agricultural area in Northwestern Europe is used by plant production. Any improvement of this utilization rate should have the highest priority with regard to energy policies. In the Middle Ages (with a cereal yield of about 8 dt/ha grain + straw), utilization rate was 0.25%. On the basis of contemporary cereal yields in Western Europe (50 dt/ha grain + 40 dt/ha straw), energetic efficiency of solar energy is 1.5%; theoretically this could be raised to 4%.

Energy input per area has increased in Western Europe's agriculture by 500% in the last century. About 40% of the total energy input is used for fuels, 30% for fertilizers and about 1 - 2% for plant protection. Energy consumption for 1 kg of a plant protection agent is compensated by a yield increase of 17.6 kg wheat (2). Experience has shown that not only energy input for chemical plant protection in cereals can be compensated many times over by a reduction of disease losses, but that with a relatively small consumption of fossil fuel used in production of modern fungicides, the utilization of solar energy by plants is considerably increased.

It has been argued that less intensive agricultural systems provide a better output/input ratio. In systems where the only input is human labor, an output/input ratio may be as high as 30:1; but this worker also consumes food containing 4.4 GJ/year. In the United Kingdom, a worker and his family use an energy level of 130 GJ/year. Thus, if the primitive agricultural system used such workers, the output/input ratio would be reduced to 1:1 compared with over 2:1 for cereals in Western Europe(6)

Therefore, the final goal can only be to improve the output/ input ratio by being very energy efficient. This can be accomplished by using plant protection measures, such as modern fungicides, to protect the energy ratio already achieved against losses by diseases and other agricultural pests.

Literature Cited

1. FAO Production Yearbook 36, 1982.
2. Green, M.B. and T. F. West, 1977: Pesticides and energy. In: Chemicals for Crop Protection and Pest Control, Oxford, pp. 23 - 28.
3. Jenkins, J.E.E., L. Lescar, 1980: Use of Foliar Fungicides on Cereals in Western Europe. Plant Disease 64, 987 - 994.
4. Kolbe, W. 1982: Weitere Versuche uber die fungizide Wirkung und den Ertragseinfluss von Baytan bei der Beizung im Getreidebau (1975 - 1981). Pflanzenschutz Nachr. Bayer 35, 72 - 103.
5. Lescar, L., 1980: Ou en est la lutte contre les maladies? Perspect. Agricoles 33, 49 - 55.
6. Lewis, D.A., J.A. Tatchell, 1979: Energy in UK Agriculture. J. Sci. Food agr. 30, 449 - 457.
7. Limburgerhof aktuell, 1984, BASF Aktiengesellschaft Landwirtschaftliche Versuchsstation Limburgerhof.
8. Priestley, R.H., R.A. Bayles, 1980: Factors Influencing Farmer' Choice of Cereal Varieties and the Use by Farmers of Varietal Diversification Schemes and Fungicides. J. Natu. Inst. Agr. Bot. 15, 215 - 230.
9. Reiner, L. et al., 1981: Weizen aktuell. DLG Verlag Frankfurt.
10. Zadoks, J.C., 1981: EPIPRE: A Disease and Pest Management System for Winter Wheat Developed in the Netherlands. EPPO Bull. 11, 365 - 369.

RECEIVED January 24, 1986

8

Fungicides and Wheat Production Technology
Advances in the Eastern United States

Herbert Cole, Jr.

Department of Plant Pathology, The Pennsylvania State University, University Park, PA 16802

Science, technology, and marketing are linked together in the broad-est sense when advances in pesticide chemistry are translated into new production systems for wheat. Science seeks to understand the wheat plant -- technology manipulates it. The level of technology used in the production of wheat is controlled by the value of the crop. When an agricultural producer contemplates various levels of input, the yield X unit price algorithm will dictate the level of technology that can be utilized in producing that crop. In the United States, wheat production has been limited as a sophisticated technological enterprise because of the linkage of U.S. wheat prices to world markets. In Europe, depending on the country, wheat prices are usually far higher than in the U.S. thus allowing a much more sophisticated level of technologic input with regard to pesticides, fertilizers, plant spacing, and permanent tramways for spraying fields. How rapidly U.S. producers emulate European production systems will depend on the U.S. wheat price.

In spite of market constraints, eastern U.S. wheat production systems are growing in complexity. No element of wheat production can be considered as an isolated unit existing in a vacuum. What can be implemented from a disease management point of view is influenced by everything else done; the use of fungicides as a technological element in the disease management component of production has become more feasible as other technological inputs are optimized. For example, the use of fungicides for disease control is dependent on successful management of other elements in the system such as new varieties, planting dates, fertilizers, application equipment, row spacing, seed spacing, and seeding rate. Overall, gains in wheat production have resulted because the whole production system has been upgraded.

This paper is an examination of the changes in wheat production that are taking place in the eastern U.S., the interaction between the various elements, and where the opportunities exist for further gains. Technological improvements have been occurring for a long time. Although minimum tillage production systems are considered a

0097-6156/86/0304-0127$06.00/0
© 1986 American Chemical Society

recent innovation, in reality farmers have been attempting to plant without plowing for at least 84 years.

Grain yield is dependent on the number of heads per hectare, the number of kernels per head and the size and density of those kernels (9). To maximize yields, every element in the whole production system must be directed towards balancing and optimizing those three contributing factors. In many ways, change in wheat production in the last ten years has been a microcosm of what has taken place in the last forty years of maize production. The changes are similar but have been compressed in time.

One of the major areas of progress has been the introduction of precision planting equipment--planters that allow depth control and precise spacing of kernels (7). Unfortunately, the goal that every seed and every plant is equidistant from every other seed or plant in the field has yet to be reached. As row spacings are narrowed from 18 cm to 10 cm, yields increase even though seeding rates have been kept constant. Much of the current development effort in planting technology is devoted in some way to achieving the equidistant seed placement.

Air seeders represent the newest way of accomplishing this goal. The concept of air seeding involves blowing a mixture of fertilizer and seed through a tube to a chisel shank that is lifting the soil. The seed is spread randomly by the air stream beneath the soil lifted by the chisel. Depth is uniform, spacing is random but remarkably equidistant. Currently manufactured seeders are suitable for the soils of the high prairies, but not for the heavy clays or stony loams of the eastern U.S. For the East the challenge is to make the system function well in heavier soils and with smaller scale equipment.

Where the air seeding concept is not functional, row planting remains a viable option. In this case, a mechanical device drops seed through tubes into rows of single or double disk openers that are forced downward to proper planting depth by the weight of the equipment. Rows preclude equidistant seed spacing, although narrow (10 cm) rows would represent a considerable advance from 18 cm rows. In attempting the ultimate goal, a wheat farmer in the state of Washington has developed a combination seeder, fertilizer and pesticide applicator that weighs approximately 25 tons. It places one herbicide over the seed row, another herbicide between the seed rows, the fertilizer beneath the seed, and then finally plants the seed, all in a precise fashion.

In the small fields of the eastern United States, different kinds of equipment may be needed. The concept of equidistant geometric seed spacing remains, but lower cost, smaller scale planters are needed to accomplish this goal. If an air seeder cannot be used in heavy soils, the alternative is to place the seed in very narrow rows using disc or chisel openers to obtain a 10 cm row spacing. Unfortunately, soil surface trash and soil clods have created problems in the successful design and operation of narrow row planters. Both small scale air and drill row planters are being used in Europe, and are now being imported into the United States. In the eastern U.S., the grain planter has often been the oldest piece of equipment on the farm. This is changing as farmers perceive new opportunities in producing wheat.

Changes in one production element may influence many other elements in the system. For example, stubble mulch fallowing every other year to conserve moisture for next year's crop had been a standard practice in certain areas of the western United States. (7). As agriculturists sought the ultimate in mechanical tillage and cropped more and more dry lands, stubble mulching was ignored and large blocks of finely prepared soil began to experience wind erosion similar to that which occurred during the dust bowls of the 1930's. In addition, people in cities and towns of the Dakotas experience an annoying side effect called SNUD, a mixture of snow and mud. After the snow melts, all surfaces are covered with this soil residue. In the dry land wheat area of the U.S., the management of soils to prevent wind erosion has become a major concern. Chemical technology is providing a large share of the solution. In Nebraska, the solution is called "ecofallow"; in other areas of the U.S. it is called chemical fallow. Regardless of name, the system involves the use of herbicides, rather than mechanical cultivation, to provide moisture retention and the control of vegetation in the intervening fallow year. This allows conservation of the soil beneath an intact plant residue and almost completely eliminates soil loss due to wind erosion.

However, in achieving the soil and moisture retention goal, another feed-back loop to a mechanical technology problem was created. How does one plant efficiently in a soil surface consisting of vegetative crop debris? Those of us in agricultural science and technology in the "intellectual centers" of the world, both corporate and university, like to think that progress occurs solely because of our great ideas. This is sometimes true, but in other instances we must rely on others. We observe and utilize unique ideas created by agricultural producers throughout the world who see solutions to thorny problems as a result of their very close association with the land. The concept of no-till or minimum tillage planting of wheat has been tried in various places throughout the world by farmers with equipment of their own creation. In one known instance, a farmer decided that he would plant wheat with his "no-till" corn planter by running it across the field offset each time so that 15-18 cm row spacing resulted. These kinds of attempts by agricultural producers throughout the world ultimately led to the development of the commercial minimum tillage or "no-tillage" drill planter. These planters helped solve an environmental problem through precise planting into soil covered by residual surface debris. Subsequent technologies allowed concurrent precise placement of the herbicide and fertilizer all in a single pass over the field. The day when the farmer used his corn planter was over.

Producers thought they had found a total solution, but it became apparent that they had then created another problem. For example, wheat plant residues contain foliar pathogens such as Septoria tritici (8). The following season, spores are produced by these fungi and are blown to adjacent fields of wheat where severe infection may result. Mechanical tillage destroys the crop residue; chemical fallow does not. So in solving one problem, another was

created. In the eastern and midwest U.S., the immediate no-tillage
planting of wheat into corn stubble, cut for ensilage, is the plant
pathologist's conceptual dream of how one inoculates a field with
Gibberella zeae to produce head scab (8). A single perithecia-con-
taining corn stalk, every 30 cm in 76 cm row spacings, will provide
enough inoculum the next spring to almost certainly ensure head scab
of the wheat.

Throughout the United States, the use of minimum tillage and
chemical fallow systems has resulted in conservation of soil and
energy, but has created a new set of disease problems that were
formerly managed by mechanical tillage and cultural practices. For
example, estimated yield reductions of up to 40% have occurred in
Pennsylvania when producers planted wheat following wheat with the
previous crop residue left on the soil surface. Major problems
resulted from Septoria diseases (Septoria nodorum and Septoria
tritici), take-all (Gaeumannomyces graminis var. tritici), head scab
(Gibberella zeae) and various other foliar, crown and root diseases.
Attempts to grow continuous wheat under minimum tillage conditions
have been discontinued by Pennsylvania farmers because of these
diseases. Similar failures have been experienced in other areas of
the eastern and mid-western U.S.

The introduction of new nitrogen responsive varieties has
greatly increased the yield potential of wheat in the areas of the
U.S. where rainfall is not a limiting factor. Older varieties merely
grew taller and lodged when additional nitrogen fertilizer was
applied. Newer varieties released in the 1978-83 period have greater
stem strength and respond to additional nitrogen by producing greater
number of tillers per hectare with larger heads. In this manner, the
basis for increased grain production has been established (4).
Unfortunately, the powdery mildew susceptibility of most wheat
varieties greatly increases as nitrogen fertility levels increase.
In seasons and locations of high humidity, many of the gains due to
increased fertility are nullified by powdery mildew losses.

Powdery mildew has long been recognized as a wheat disease, but
only with the advent of high nitrogen fertility did it become a major
yield limiting factor in wheat production. Fungicides for powdery
mildew control have evolved slowly. Sulfur was the first product and
it has been available since the early 1900's, but it has never been
effective enough for commercial use. Then came ethirimol, (5 Butyl-
2-ethylamino-4-hydroxy-6-methyl pyrimidine), a compound proven to
provide effective powdery mildew control in Europe, but be only
marginally effective in the U.S., and was never registered for
commercial use. Triadimefon (1-(4-chlorophenoxy)-3,3-dimethyl-1-
(1H-1,2,-4,-triazol-1-yl)-2-butanone) was the first EPA registered
commercially effective powdery mildew fungicide. The use of triadimefon on high fertility responsive, powdery mildew susceptible
varieties has resulted in yield increases of 10-30 percent in absence
of other foliar pathogens.

It is unfortunate, but not surprising, that a fungicide effective only against powdery mildew, such as ethirimol, allows other
pathogens not sensitive to the fungicide to invade the unaffected
foliage. For example, where powdery mildew had been the primary
problem, Septoria tritici and Puccinia recondita now invaded the

foliage from which powdery mildew had been eliminated (3). Thus, technological innovation solved one problem, but again created a new problem. It became apparent that an additional fungicide must be added to the spray tank to control Septoria leaf blotch. An application method also had to be developed. One possible solution was aerial spraying; the other was the development of ways to ground spray in mid-season without excessive plant damage. By developing narrow wheel truck or tractor mounted sprayers, smaller fields could be sprayed from the ground with very little damage. The wheel tracks represented an acceptable level of yield loss (1-2%).

Additional new systemic fungicides are being introduced for wheat disease control. One material CGA 64250 (Tilt),(1-(2-(2,4-dichlorophenyl)4-propyl-1,3,-dioxolan-2-yl methyl)-H-1,2,4-triazole), is effective for control of powdery mildew, rust, and Septoria leaf blotch. However at this point in time, it is not certain whether U.S. EPA registration will allow a short enough interval between treatment and harvest for effective disease control by this compound.

Foliar sprays require application during the growing season. An alternative would be seed treatment with a long-term systemic fungicide that could be applied at planting. Such a material is currently being tested. Triadimenol (Baytan),(β-(4-Chlorophenoxy)-α-(1,1-dimethylethyl)-1H-1,2,4-triazole-1-ethanol), when applied as a seed treatment, will provide control of powdery mildew on wheat into mid-season. In most years, this would preclude the need for foliar sprays for powdery mildew control, but does not provide a solution for control of Septoria leaf blotch or late season leaf rust.

The new systemic fungicides offer a great potential benefit; however, the mechanisms of fungicidal action of these products tend to be narrow, often based on inactivation of a single enzyme system. With wide usage of these fungicides, pathogen resistance may become a major problem. The natural mutation rate of the fungal pathogen combined with almost universal exposure to the fungicide provides a basis for selection of resistant strains of the pathogen. If these strains are fit to survive and compete in the host/pathogen ecosystem, the fungicide will have a very short life expectancy. Benomyl was hailed as the solution to many plant diseases. However, resistance has occurred in Cercospora, Erysiphe, Botrytis, Sclerotinia, Venturia, and other fungal genera. Fortunately, resistance to triadimefon has not yet appeared in wheat pathogens in areas of the world where the fungicide has been in widespread use for at least ten years. This is a hopeful sign. Triadimenol is closely related to tridimefon but has not yet been used extensively and one can only speculate regarding future resistance problems.

In a disease control experiment in 1979 (H. Cole, unpublished), triadimefon was applied to four wheat varieties ('Red Coat', 'Hart', 'S-76', and 'Dancer') grown under both a high fertility regime conducive to maximum grain yields, and the standard moderate fertility regime. In the case of 'Red Coat', a powdery mildew resistant variety, the high fertility resulted in very little yield gain (3555 kg/ha vs 3230 kg/ha); yield responsiveness to fertility was not present in the variety. On the other hand, 'Hart' and 'S-76' responded dramatically to high fertility. The use of triadimefon under standard fertility resulted in a 20% yield increase, whereas the increase was almost 70% under high fertility plus fungicide. High fertility alone resulted in no yield increase due to the extremely severe powdery mildew levels that were experienced. The maximum yields obtained from the fungicide plus high fertility treatment plots were 6370 kg/ha for 'Hart' and 6150 kg/ha for 'S-76'. The variety 'Dancer', which is extremely susceptible to powdery mildew, produced a poor yield (2340 kg/ha) at moderate fertility in the absence of powdery mildew control, and with the use of triadimefon produced a moderate grain yield of 3430 kg/ha. The combination of high fertility plus triadimefon on 'Dancer' resulted in only a slight yield increase (7%) because the yield responsiveness to fertility was not present in this variety. These results and many others clearly indicate that yield potential and fertilizer responsiveness must be present in a variety in order for the use of fungicides to result in dramatic yield increases.

We have also examined the interactions between fumigation, to control root pathogens, and fungicide sprays, to control foliar pathogens. This combination created even greater complexities. The soil fumigation treatment consisted of 400 kg ai/ha of methyl bromide applied three weeks prior to planting winter wheats. The foliar spray was a combination of triadimefon (140 g ai/ha) and mancozeb (1 kg ai/ha) applied at Feekes' growth stages 7 and 10 (5). The experimental design was factorial so that both fumigation and foliar fungicide application could be considered alone as well as together. Depending on variety, there were some instances where yield increases resulted from foliar sprays alone and fumigation alone. In other instances, the combination treatment resulted in additive yield increases and in other instances synergistic increases greater than additive. We do not understand the interrelationships among pathogens, especially root and crown rot pathogens versus the foliar pathogens. We are certain that there is an interaction and that it must be considered in development of wheat production technologies.

The one thing that I have learned over the years in examination of data is my own tendency to expect certain kinds of results. If they are not obtained, I try to explain this by blaming the experiment and blaming everything but the fact that I may really be looking at a unique biological paradox that needs to be understood rather than quarreled with. It has been said that the difference between "genius" and the "rest of us" is that "genius" sees relationships between seemingly unrelated observations when the "rest of us" cannot see these relationships.

Fungicides cannot be developed and tested in isolation from the other technological elements in a wheat production system. Where researchers have discovered an effective new fungicide and then sprayed it on test plots without regard to the whole system, they have failed miserably. A chemical company several years ago attempted to do that with a product that had been effective in Europe. They sprayed wheat from Florida to New York, spent several million dollars, and went home totally disenchanted. They ignored the fact that fungicides are a part of a complex system that involves more than merely spraying a chemical on a crop.

The one element that has not been discussed is that currently popular term "biotechnology". To most people, the word means gene splicing and genetic engineering by chemical or artificial means. This is unfortunate since biotechnology in the holistic sense involves all manipulation of biological systems, including classic plant breeding. The term is not as restrictive as its current usage would imply. There has been a great deal of effort by university researchers, as well as by private seed companies, to develop wheat varieties that will maximize grain yields. Hybrid wheats have been a long-time goal but to date have achieved only limited commercial success. The large degree of self-pollination and the absence of easily manipulated male sterility factors have made hybridization a difficult commercial process (1,2). Gameticidal chemicals are currently receiving research emphasis. However, the vast majority of available varieties have been developed by classic plant breeding techniques with major emphasis devoted to straw strength, disease resistance, fertilizer responsiveness, and yield (6).

Long lasting resistance would be the ideal solution to many plant diseases. Unfortunately, the narrow genetic basis of the resistance, often a single dominant gene, has resulted in new races of the pathogen which bypass the resistance gene after several years of widespread use of the new variety. In some instances, such as with 'Tyler', a new soft red winter wheat variety, the benefits of powdery mildew resistance may be totally offset by leaf rust susceptibility. Although there has been a measurable long-term benefit from breeding for disease resistance, progress to date has not eliminated the need for use of fungicides. Genetic engineering and gene splicing may ultimately benefit wheat production but no progress has yet occurred. "Super wheat" has not yet appeared and all things considered, will probably never come about.

From the preceding discussion, it is apparent that we cannot consider any technological practice in a vacuum. Wheat production is a system composed of many interacting elements in which every element influences other elements. One could draw the analogy that it is the same as pressing an indentation in the side of a plastic bag of water--a push in one place will guarantee that there will be an equal volume pushed out somewhere else. Where it will appear is not readily determined but the fact that it will appear is assured. The same is true in attempting to understand the interrelationships in technology. The major challenge that we face today is to recognize and deal with the issues of technological change in a holistic sense. Unfortunately, we much prefer to zero in on one's own specific interest, pursue research in that direction and not worry about the side effects or ramifications. We should be encouraging people to examine systems of crop production, and to examine technologies as they interact rather than as individual elements. I believe that new useful technologies will be developed and incorporated into the wheat production systems and that fungicides will play an increasing role in this development. The objective for all of us should be to ensure that we make real progress, especially in profitability and yield increase, so that wheat producers do more than just spend more money in production costs without at least a slight gain in profit.

Literature Cited

1. Ausemus, E.R.; F.H. McNeal; J.W. Schmidt. 1967. Genetics and
    Inheritance. Pages 225-259 In "Wheat and Wheat Improvement";
    Quisenbery, K.S. and L.P. Reitz, Eds.; Agronomy Monograph No. 13
    Amer. Soc. Agron. Madison, Wisconsin.
2. Briggle, L.W. 1963. Heterosis in Wheat -- A Review. Crop Sci.
    3: 407-412.
3. James, W.C. and C.S. Shih. 1973. Relationship between Inci-
    dence and Severity of Powdery Mildew and Leaf Rust on Winter
    Wheat. Phytopathology 63:183-187.
4. Laloux, R., A. Falisse, and J. Poelaert. 1980. Nutrition and
    Fertilization of Wheat. Pages 19-24 In "Wheat Documenta.";
    Technical Monograph, Ciba Geigy L&D, Basle, Switzerland.
5. Large, E.C. 1954. Growth Stages in Cereals. Plant Pathology
    3:128-129.
6. Poehlman, J.M. 1959. "Breeding Field Crops" Chapter 6. Henry
    Holt and Co., Inc., New York.
7. Schlehuber, A.M. and B.B. Tucker. 1967. Culture of Wheat,
    Pages 117-176 In "Wheat and Wheat Improvement"; K.S. Quisen-
    berry and L.P. Reitz, Eds., Agronomy Monograph, No. 13, Amer.
    Soc. Agron., Madison, Wisconsin.
8. Wiese, M.V. 1977. "Compendium of Wheat Diseases"; The Ameri-
    can Phytopathological Society, St. Paul, Minn.
9. Willey, R.W. and S.B. Heath. 1964. The Quantitative Relation-
    ship between Plant Population and Yield. Adv. Agron. 21:
    281-320.

RECEIVED October 1, 1985

Tree Fruit Crops in the Eastern United States
Potential Role for New Fungicides

Alan L. Jones

Department of Botany and Plant Pathology, Michigan State University, East Lansing, MI 48824

Fungicides are essential for the production of high quality deciduous tree fruit crops in the humid eastern United States. On apples, ten to fifteen applications of one to three fungicides are used each season. Although the frequent application of fungicides is costly, dependence upon chemicals for the production of disease-free fruit crops will continue for the foreseeable future.

A number of experimental fungicides that have the potential for easing the complexity of disease control on tree fruit crops in the east are currently being evaluated. Most of these fungicides are either ergosterol biosynthesis inhibiting (EBI) compounds or members of the dicarboximide group of fungicides. Research has been conducted at several experiment stations throughout the eastern United States to determine the effectiveness of these fungicides against the spectrum of diseases in the various regions and to define application rates. In addition, specific studies have been conducted to establish the practical mode of action of these compounds for controlling the most important diseases.

Activity Against Apple Scab

Because the management of apple diseases revolve around control strategies for apple scab (Venturia inaequalis), specific studies have been conducted on the capabilities and limitations of the EBI fungicides to control this disease. The EBI fungicides inhibit the spore infection process at a later stage than earlier fungicides. Fungicides such as captan, metiram and sulfur inhibit spore germination while the benzimidazole group of fungicides inhibit appressoria formation but not spore germination. With the EBI fungicides, spores germinate, form appressoria, and penetrate the cuticle of the host before they are killed. This is exemplified in an experiment where an apple leaf was treated with fenarimol 12 hours before inoculation with conidia of Venturia inaequalis and then incubated in a moist chamber for 23 hours after inoculation (Figure 1). Although the spore produced a germ tube, appressorium, and penetrated the cuticle within 24 hours after inoculation, no scab symptoms developed on

treated plants after two weeks, while unsprayed plants were severely
infected. Conidia on plants treated with bitertanol and etaconazole
reacted similarly, and no disease symptoms were observed after two
weeks.

Greenhouse studies indicate that EBI fungicides provide about
one to four days of protection compared to five to seven days for
conventional protectants such as captan, metiram, and mancozeb (20,
23). The degree of protection by the EBI fungicides depends on
dosage, on the specific fungicide used, on the weather conditions
(particularly rain), and on the rate of plant growth after the
application. Mixtures of EBI fungicides with conventional fungicides
are currently being evaluated as a means of improving the protective
action and spectrum of activity of the EBI fungicides.

The EBI fungicides have outstanding after-infection control
activity compared to conventional fungicides. The limit of after-
infection control for a particular fungicide is defined as the number
of hours from the start of a wet period that a spray can be applied
and still prevent development of visible symptoms. Conventional
fungicides perform best when applied in anticipation of infection
periods. Their effectiveness is very limited when they are applied
after conditions are favorable for infection. Fungicides such as
captan, metiram, and mancozeb are effective for 18 to 24 hours after
the initiation of favorable conditions for infection while the EBI
fungicides are effective for 72-96 hours (12,19,20,23). The specific
duration of after-infection activity is proportional to the amount of
active ingredient that is applied (20) and, in the specific case of
bitertanol, it can be improved by adjuvant addition (19).

When the EBI fungicides are applied later than 72-96 hours after
the beginning of the scab infection period but before symptoms are
visible, chlorotic spots or flecks with few or no conidia are produced in place of typical sporulating lesions (Figure 2). This is
referred to as presymptom control by Szkolnik (23) and it is a form
of post-infection or curative action. Dodine and benomyl are examples of fungicides with presymptom activity. Presymptom activity
becomes more obvious as the time after inoculation increases beyond the
72-96 hour after-infection period. In greenhouse trials, sprays of
bitertanol, etaconazole, fenarimol and triforine reduced conidial
production 85 to 99% when applied five days after inoculation (20).

Presymptom activity has also been observed in field trials. If
sprays are delayed late into the incubation period, presymptom
control can be increased by applying a second spray seven days after
the first application (12). There are indications that chlorotic
lesions will not produce conidia after the use of EBI fungicides is
discontinued (6). However, the viability of the fungus in these
lesions is influenced by moisture, dosage of fungicide, timing, and
other factors. In one trial, the scab fungus was isolated from
chlorotic lesions in leaves (12).

Because of their after-infection and antisporulation activity
against scab, the EBI fungicides are especially suitable for use in
predictive disease control programs. Predictive disease control is
an important part of integrated pest management programs on apples.
A predictive system means control measures are applied after the
onset of infection. Small special purpose computers with field

Figure 1. A germinated conidium of <u>Venturia inaequalis</u> on a apple leaf treated with fenarimol 12 hours before inoculation and incubated in a moist chamber for 23 hours after inoculation. Bar = 1 micron.

Figure 2. Densely sporulating lesions of apple scab on an unsprayed leaf (left) compared with chlorotic nonsporulating lesions on a leaf sprayed with etaconazole (right).

sensors can be utilized as an aid in identifying infection periods
and timing fungicide sprays (11). Because they are limited in
protective action, the most efficient use of the EBI fungicides is in
conjunction with a predictive system. Recently, the value of a
microcomputer to enhance the efficiency of bitertanol, fenarimol,
etaconazole, and triforine was demonstrated under field conditions
(2).
        Trials with EBI fungicides used as post-symptom treatments
indicate they are less effective than dodine and benomyl for stopping
the development of established lesions. In greenhouse trials involving the removal of conidia from sporulating lesions and spraying the
lesions immediately with EBI fungicides, numbers of conidia were not
reduced after three days (23). However, in field trials involving
two sprays of bitertanol or etaconazole applied seven-days-apart
starting about two days after symptoms were visible, immature spores
predominated in the lesions seven days later (12). The immature
spores were difficult to remove from the lesions and those that were
removed failed to germinate. Failure to obtain a reduction in number
of conidia in greenhouse trials, but not in field trials involving
multiple sprays, may be related to the duration of time needed to
deplete the reserves of ergosterol within the fungus. On fruit,
post-symptom control was exhibited by a "healing" of the scab lesions
(12).
        The post-infection properties of these fungicides suggest they
are systemic in apple foliage. Limited studies with etaconazole and
fenapanil indicate there is foliar uptake or at least a binding of
the fungicide to the cuticle (13). Additional research on the rate
of uptake and persistance of these compounds may help to resolve the
practical question of how long spray deposits must dry before they
reach maximum effectiveness. Field observations suggest that a rain
during or immediately after the application of an EBI fungicide
reduces its effectiveness.
        The importance of the EBI fungicides is increasing because of
the widespread development of resistance in the scab fungus to dodine
and/or the benzimidazole fungicides in the eastern United States
(4,8,25). The use of these older fungicides have been discontinued
in some areas, forcing growers to use less effective fungicides and
more total chemical per hectare than was previously required. The
potential for development of resistance in the scab fungus to the EBI
fungicides may be of greatest concern in the northern states. In
this region the spectrum of diseases controlled by the EBI fungicides
is sufficient to permit season long programs. In southern areas,
other fungicides must be mixed with the EBI fungicides and this
should help delay resistance problems.

Activity Against Other Apple Diseases

A large number of fungal diseases in addition to scab affect apples
in the mid-Atlantic and southern states. In areas within these
states, the following diseases must be controlled: cedar apple rust
(Gymnosporangium juniperi-virginianae), quince rust (G. clavipes),
powdery mildew (Podosphaera leucotricha), black rot (Physalospora
obtusa), bot or white rot (Botryosphaeria dothidea), bitter rot

(<u>Glomerella</u> <u>cingulata</u>), sooty blotch (<u>Gloeodes</u> <u>pomigena</u>), fly speck (<u>Microthyriella</u> <u>rubi</u>) and Brooks spot (<u>Mycosphaerella</u> <u>pomi</u>).

The EBI fungicides have shown outstanding activity against apple powdery mildew everywhere in the eastern United States. Mildew control was improved particularly in the mid-Atlantic region (<u>5</u>,<u>29</u>), where the disease is more severe than in the northeastern and north central states. The EBI fungicides differ from older mildewcides like sulfur, dinocap, and benomyl by inhibiting sporulation and mycelial growth in established lesions (<u>5</u>). Their high level of control efficiency is probably due to a reduction in both primary inoculum and secondary spread. As with dinocap (<u>29</u>), the EBI fungicides may be more effective when applied on a 5- to 7-day interval at low rates rather than on a 10- to 14-day interval at high rates. Substantial control of powdery mildew has also been obtained in greenhouse trials through vapor action with triadimefon and etaconazole (<u>23</u>). The role of vapor action in controlling mildew in orchards remains unclear.

Cedar apple rust is a critical problem where susceptible apple cultivars are grown in proximity to cedar trees. The disease is a problem in part of New York state and most of the southern and north central states. Use of protectant fungicides has been the only method available to control infection in areas where eradication of red cedar trees is impractical. The EBI fungicides, except prochloraz, have shown better activity against cedar apple rust than standard protectant compounds (<u>27</u>). Dosages needed to control rust are generally lower than those used to control scab and mildew (<u>27</u>). The capability of the EBI fungicides to give post-infection control of cedar apple rust should result in the development of new control strategies for this disease (<u>15</u>,<u>22</u>).

Although the EBI fungicides are of greatest benefit for the control of scab, rust, and powdery mildew early in the season, their benefit in the cover spray period will differ between production areas. In the northern United States where apple scab and powdery mildew are the main diseases to be controlled, the EBI fungicides will be of benefit from the beginning of the growing season until harvest. However, in the southern, mid-Atlantic, and southeastern states, a complex of summer fruit rot diseases is common. Because the EBI fungicides are weak against this group of diseases, they will need to be mixed with more effective fungicides such as mancozeb or captan (<u>27</u>). Bitertanol should fit better into mid- to late-season programs than the other EBI fungicides because of its stronger activity against sooty blotch and fly speck (<u>28</u>).

<u>New Fungicides for Stone Fruit Disease Control</u>

Fungicides are used on stone fruit crops in the eastern United States to control brown rot blossom blight and fruit rot (<u>Monilinia</u> <u>fructicola</u>), cherry leaf spot (<u>Coccomyces</u> <u>hiemalis</u>), powdery mildews (<u>Sphaerotheca</u> <u>pannosa</u> and <u>Podosphaera</u> <u>oxyacanthae</u>), and scab (<u>Cladosporium</u> <u>carpophilium</u>). In the northeastern states about six applications are applied per season while in the southeastern states up to ten applications are applied.

The need for new fungicides for stone fruits has increased because of the development of benzimidazole-resistant Monilinia fructicola (8,9,30) and Coccomyces hiemalis (10) in important stone fruit production areas of the eastern United States. Resistance has been a major problem for stone fruit growers because alternative fungicides are less effective, more costly or require more frequent applications than benzimidazole fungicides. The recent registrations of iprodione and triforine for control of stone fruit diseases have helped to alleviate problems caused by benzimidazole resistance.

Programs for the management of stone fruit diseases revolve around the control of brown rot at bloom and at harvest. Between these two stages of fruit development, the control of cherry leaf spot, powdery mildew, and peach scab is more important than control of brown rot. The EBI fungicides have shown high activity against all these diseases except peach scab. Peach scab is a mid-season problem in the southern states. In these states the EBI fungicides would need to be supplemented with other fungicides for scab control. In the northern states, the EBI fungicides would be of benefit for controlling the spectrum of diseases that occur from bloom until harvest.

Cherry leaf spot is particularly severe in the Great Lakes states of Michigan, Wisconsin, and New York. Protective fungicides such as captafol, chlorothalonil, captan, and ferbam plus sulfur are the main fungicides used in its control. These fungicides lack the broad spectrum activity of the EBI fungicides and often must be supplemented with other fungicides to control all the diseases that are present. Also, the capability of the EBI fungicides to give post-infection control of leaf spot should be a valuable asset in the control of this disease (21). If protective sprays are missed and leaf spot infection occurs, the EBI fungicides could be used as emergency treatments in much the same way that cycloheximide was used, before its use was discontinued (3).

A second group of fungicides with potential for use on stone fruits is the dicarboximides. The dicarboximide fungicides provided excellent control of brown rot blossom blight in laboratory tests in Michigan (7) and both dicarboximide and EBI fungicides provided excellent control of brown rot blossom blight when applied 18 or 24 hours after inoculation in greenhouse trials in New York (24). Because of this after-infection activity, the dicarboximide and the EBI fungicides should be substituted for dichlone in emergency control programs for brown rot blossom blight. For this purpose, the fungicides are applied after conditions have been favorable for infection. The dicarboximides have a narrower spectrum of disease control than the EBI fungicides. At times, they will need to be supplemented with other fungicides to control the entire spectrum of diseases that is present.

Laboratory tests with M. fructicola (26) and field experience with the dicarboximides for Botrytis control in strawberry fields (14) suggest fungicide resistance may be a problem if the dicarboximide fungicides are used extensively and continuously season after season. Although many workers consider dicarboximide resistant strains less fit than wild-type strains (16,17), the research suggests that field resistance will develop if steps are not taken to avoid the problem. As pointed out by Gilpatrick (4), the main benefit of having a wide variety of fungicides available for brown rot control is to prevent the development of resistance. Another benefit of having a wide variety of fungicides available for stone fruits would be the possibility of controlling a broader range of diseases.

A Need for Root Rot Fungicides

Phytophthora crown rot, caused by Phytophthora cactorum and other Phytophthora spp., occurs on apple rootstocks such as MM104 and MM106 planted in heavy poorly drained soils. Soil drenches with copper sulfate, captafol, and mancozeb are not highly effective for controlling crown rot, but metalaxyl has been effective in limited trials (1). Fosetyl-Al is also active against Phytophthora, but it has not yet been investigated extensively for use in the eastern United States. The potential value of these compounds is uncertain because growers often fail to treat trees until the infection has progressed beyond a stage where the tree can recover. Moreover, through proper rootstock and site selection, crown rot can be pre-vented.

A Need for Antibiotics

Since the era of the antibiotics, no new compounds have been developed for the control of diseases caused by phytopathogenic prokaryotes. Even these antibiotics are not used in many areas of the world because of governmental restrictions that prevent their use on agricultural crops. Problems with resistance are reducing the usefulness of the antibiotics even further. Copper compounds, which are generally less effective than the antibiotics, are the main compounds available to control bacterial incited diseases. Unfortunately, copper compounds are phytotoxic to many crops.

The need for effective bactericides is greater today than at any time in history. The recognition that mycoplasma-like organisms and xylem-limited bacteria cause plant disease means that there are additional diseases that are amenable to control with antibiotic-like compounds. X-disease, pear decline, peach yellows, phony peach, and plum leaf scorch are a few examples of diseases on deciduous tree-fruit crops caused by phytopathogenic prokaryotes.

Highly effective chemicals are also needed to control bacterial diseases of tree fruit and many other high value crops. Current methods for controlling fire blight, bacterial spot, and bacterial canker are not adequate. Fire blight, for example, has spread to Europe. The role of Pseudomonas syringae pv. syringae as an "ice nucleator" has stimulated the hunt for frost protection compounds that control or compete with these bacteria. New chemicals for the control of phytopathogenic prokaryotes are also needed as tools for avoiding resistance problems. Industry should recognize that the potential market for a new chemical cannot be estimated accurately based on current sales of available compounds (18).

Summary
For many years virtually no fungicides with new chemistry were
approved for use on tree fruit crops in the United States because of
the slow federal registration process. Many experimental fungicides
have been evaluated by agricultural experiment stations in several
states in the eastern United States during the last ten to fifteen
years. At least three of these new fungicides (triforine, triadimefon, and iprodione) are now registered for use on tree fruit crops,
but many additional fungicides are awaiting final federal registration. As these new registrations are approved, the potential promise
of these fungicides to improve disease pest management on tree fruit
crops can finally be realized.

Literature Cited
1.   Ellis, M.A., Grove, G.G., and Ferree, D.C. 1982. Effects of
     metalaxyl on Phytophthora cactorum and collar rot of apple.
     Phytopathology 72:1431-1434.
2.   Ellis, M.A., Madden, L.V., and Wilson, L.L. 1984. Evaluation
     of an electronic apple scab predictor for scheduling fungicides
     with curative activity. Plant Dis. 68:1055-1057.
3.   Eisensmith, S.P., and Jones, A.L. 1981. Infection model for
     timing fungicide applications to control cherry leaf spot.
     Plant Dis. 65:955-958.
4.   Gilpatrick, J.D. 1983. Management of resistance in plant
     pathogens. Pages 735-767 In: "Pest Resistance to Pesticides".
     G.P. Georghiou and T. Saito, eds. Plenum Press, N.Y. 809 pp.
5.   Hickey, K.D., and Yoder, K.S. 1981. Field Performance of
     sterol-inhibiting fungicides against apple powdery mildew in the
     mid-Atlantic apple growing region. Plant Dis. 65:1002-1006.
6.   Hoch, H.C., and Szkolnik, M. 1979. Viability of Venturia
     inaequalis in chlorotic flecks resulting from fungicide applica-
     tion to infected Malus leaves. Phytopathology 69:456-462.
7.   Jones, A.L. 1975. Control of brown rot of cherry with a new
     hydantoin fungicide and with selected fungicide mixtures. Plant
     Dis. Rep. 59:127-130.
8.   Jones, A.L. 1981. Fungicide resistance: Past experience with
     benomyl and dodine and future concerns with sterol inhibitors.
     Plant Dis. 65:990-992.
9.   Jones, A.L., and Ehret, G.R. 1976. Isolation and characteriza-
     tion of benomyl-tolerant strains of Monilinia fructicola. Plant
     Dis. Rep. 60:765-769.
10. Jones, A.L., and Ehret, G.R. 1980. Resistance of Coccomyces
     hiemalis to benzimidazole fungicides. Plant Dis. 64:767-769.
11. Jones, A.L., Lillevik, S.L., Fisher, P.D., and Stebbins, T.C.
     1980. A microcomputer-based instrument to predict primary apple
     scab infection periods. Plant Dis. 64:69-72.
12. Kelley, R.D., and Jones, A.L. 1981. Evaluation of two triazole
     fungicides for post-infection control of apple scab. Phytopath-
     ology 71:737-742.
13. Kelley, R.D., and Jones, A.L. 1982. Volatility and systemic
     properties of etaconazole and fenapanil in apple. Can. J. Plant
     Pathol. 4:243-246.

14. Maraite, H., Gills, G., Meunier, S., Weynes, J., and Bal, E. 1980. Resistance of Botrytis cinerea Per. ex Pers. to dicarboximide fungicides in strawberry fields. Parasition 36:90-91.
15. Pearson, R.C. Szkolnik, M., and Meyer, F.W. 1978. Suppression of cedar apple rust pycnia on apple leaves following post-infection applications of fenarimol and triforine. Phytopathology 68:1805-1809.
16. Ritchie, D.F. 1983. Mycelial growth, peach fruit-rotting capability, and sporulation of strains of Monilinia fructicola resistant to dichloran, iprodione, procymidone, and vinclozolin. Phytopathology 73:44-47.
17. Ritchie, D.F. 1982. Effect of dichloran, iprodione, procymidone, and vinclozolin on the mycelial growth, sporulation, and isolation of resistant strains of Monilinia fructicola. Plant Dis. 66:484-486.
18. Schroth, M.N., and McCain, A.H. 1981. The nature, mode of action, and toxicity of bactericides. In "Handbook of Pest Management in Agriculture" (D. Pimental, ed.) Vol. III, pp. 47-58. CRC, Cleveland, Ohio.
19. Schwabe, W.F.S., and Jones, A.L. 1983. Apple scab control with bitertanol as influenced by adjuvant addition. Plant Dis. 67:1371-1373.
20. Schwabe, W.F.S., Jones, A.L., and Jonker, J.P. 1984. Greenhouse evaluation of the curative and protective action of sterol-inhibiting fungicides against apple scab. Phytopathology 74:249-252.
21. Szkolnik, M. 1974. Unusual post-infection activity of a piperazine derivative fungicide for the control of cherry leaf spot. Plant Dis. Rep. 58:326-329.
22. Szkolnik, M. 1974. Unique post-infection control of cedar-apple rust on apple with triforine. Plant Dis. Rep. 58:587-590.
23. Szkolnik, M. 1981. Physical modes of action of sterol-inhibiting fungicides against apple diseases. Plant Dis. 65:981-985.
24. Szkolnik, M. 1983. Greenhouse evaluations on protective and after-infection activities of fungicides in the control of sweet cherry brown rot blossom blight. Fungicide and Nematicide Tests 38:117.
25. Szkolnik, M., and Gilpatrick, J.D. 1969. Apparent resistance of Venturia inaequalis to dodine in New York apple orchards. Plant Dis. Rep. 53:861-864.
26. Sztejnberg, A., and Jones, A.L. 1978. Tolerance of the brown rot fungus Monilinia fructicola to iprodione, vinclozolin and procymidone fungicides. (Abstr.) Phytopathol. News 12:187.
27. Yoder, K.S., and Hickey, K.D. 1981. Sterol-inhibiting fungicides for control of certain diseases of apple in the Cumberland-Shenandoah region. Plant Dis. 65:998-1001.
28. Yoder, K.S. 1982. Broad spectrum apple disease control with bitertanol. Plant Dis. 66:580-583.
29. Yoder, K.S., and Hickey, K.D. 1983. Control of apple powdery mildew in the mid-Atlantic region. Plant Dis. 67:245-248.
30. Zehr, E.I. 1982 Control of brown rot in peach orchards. Plant Dis. 66:1101-1105.

RECEIVED October 1, 1985

Fungicides for Disease Control in Grapes
Advances in Development

Roger C. Pearson

Department of Plant Pathology, New York State Agricultural Experiment Station, Cornell
University, Geneva, NY 14456

The grape is the most widely planted fruit crop in the world. Spain,
Italy, Russia and France, each with over one million hectares of
grapes, easily outrank the United States (US) which has 323,000
hectares. As of 1981, the worldwide planting of grapes totaled over
ten million hectares (Table I) (1). The US ranks sixth among the
wine producing countries (Table II) and first among the raisin
producing countries. In 1982, the commodity value of grapes for
fresh and processing uses in the US was $1.34 billion, second only to
potatoes for all fruits and vegetables (4).

Table I. The World's Vineyards in 1981.

Country	Hectares (1,000)
Spain	1,650
Italy	1,360
U.S.S.R.	1,353
France	1,154
Turkey	801
Portugal	360
Argentina	324
U.S.A.	323
Romania	307
Other countries	2,441
TOTAL	10,073

Adapted from Ref. 1.

    Currently, there is a worldwide wine glut which is the result of
several factors, among them bumper crops of 1982 and 1983 coupled
with a steady decline in consumption in the major European wine
producing countries (Table II) (16). Due to the current strength of
the US dollar, American wine is placed in a noncompetitive position
in the world market. For example, a Bank of America study (14)
showed that in 1979, when the dollar was weak, US wine exports grew
almost 120% while imports declined 2%. In 1982, when the dollar was
strong, imports grew 6% while exports dropped 11%.

0097–6156/86/0304–0145$06.00/0
© 1986 American Chemical Society

Table II. Wine Consumption and Production in Various Countries.

Country	Per Capita Consumption (Liters) 1970	1982	Production, 1982 (1,000 hectoliters)
France	109	86	79.2
Italy	111	83	72.6
Spain	62	57	37.3
U.S.S.R.	11	14	34.6
Argentina	92	74	25.0
U.S.A.	5	8	19.5
West Germany	16	25	15.4
Portugal	77	78	10.0
South Africa	11	10	8.9
Romania	23	29	8.7
Yugoslavia	27	28	8.6
Hungary	38	30	6.8
Chile	44	55	6.1
Greece	40	44	5.5
Austria	38	35	4.9
Bulgaria	19	22	4.9
Australia	9	19	4.0
Denmark	6	16	–
Netherlands	5	13	–
U.K.	3	7	–

Adapted from Ref. 16.

The wine glut is also partially to blame for the current huge raisin surplus in the US. As young plantings of wine varietals have come into bearing in California and wine sales have slumped, wineries have reduced purchases of Thompson Seedless grapes, forcing many of the Thompson Seedless growers to make raisins. Furthermore, the entry of Greece, Portugal and Spain into the European Economic Community has eroded the primary export market for California raisins, that of northern Europe.

Despite the current bleak picture for the wine industry, the romantic appeal of grapes and wine making continues to lure people into the business. For example from 1974 to 1983, bonded wineries in the US increased from 510 to 1,114. Since 1980, 292 new wineries have been licensed (15). Currently there are bonded wineries in 41 states; the latest to be added was in the state of Maine.

Although the short-range outlook for the wine industry is bleak, the long-term outlook is bright according to California's Bank of America (14), which predicts that wine consumption will continue to expand in the US at an average annual rate of 6%. Americans will choose to drink more wine as the economy improves, the population in the 25- to 45-year age group increases, states relax regulation of alcohol sales, and as marketing campaigns persuade consumers to drink more wine. Currently, 10% of the US population consumes 66% of all the wine utilized in the US. The bank predicted that by 1990, wine production will need to rise to allow for a projected 45% increase in sales.

One bright spot in the grape industry is the table grape market. The per capita consumption of table grapes in the US increased from 0.8 kg in 1972 to 2.6 kg in 1982, resulting in an annual increase of

20% per year (4). The per capita consumption of fresh fruit places
grapes fourth behind bananas, apples, and oranges. Apparently,
increased consumer emphasis on natural, nutritional, and low calorie
foods has had a favorable impact on the table grape market.

In the US grape industry, California is far ahead of other
states with 92% of the acreage and 93% of the production as of 1982.
Sales of wine made in states other than California account for less
than 10% of the US wine sales. Furthermore, California produces
almost 100% of the raisins and nearly that amount of the table grapes
grown in the US. New York State, for example, has approximately
17,000 hectares of grapes and is a distant second to California.

The position of the US grape industry, particularly its current
economic depression, should be kept in perspective when recent developments in fungicides and their impact on the industry are discussed.
First, it would be useful to review the major fungal diseases of
grapes in the US. These include four fungal diseases in eastern
vineyards, two of which also occur in California vineyards.

Powdery mildew (Uncinula necator (Schw.) Burr), or Oidium as it
is called in Europe, is perhaps the most common fungal disease of
grapes worldwide. This fungus attacks all green parts of the vine,
but is most obvious on leaves and fruit where it appears as a white
to grayish-white dusty covering. When leaves become infected while
unfolding or expanding, their development is interrupted and they
become curled and distorted. When berries are infected at an early
stage in their development, they either abort or the epidermis of the
berry stops growing and, as the pulp continues to expand, the berry
cracks. Diseased fruit account for considerable crop loss and
cracked fruit are susceptible to rot organisms. Wine made from
infected fruit has an objectionable off-flavor. Since the fungus
develops under a wide range of humidity and temperature conditions
and is not favored by rainfall, distinct infection periods are
difficult to identify.

Downy mildew, caused by Plasmopara viticola (Berk. & Curt.)
Berl. & de Toni, occurs only in climates that are wet during the
growing season and therefore not in California. The fungus can
infect all green parts of the vine that have functional stomata.
Infections on leaves first appear on the upper surfaces as oily spots
that quickly develop into yellowish lesions, the undersides of which
are covered with white sporangia that spread the disease. When young
shoots, petioles, or tendrils become infected, they become distorted
and curled. Major crop loss can occur when fruit become diseased.
Infection periods are identified on the basis of temperature and
duration of leaf wetness.

Black rot, caused by Guignardia bidwellii (Ellis) Viala and
Ravaz, is favored by warm humid weather. Rainfall is necessary for
disease buildup and spread, eliminating occurrence in California.
Black rot is characterized by brown circular lesions which may co-
alesce to destroy large areas of diseased leaves or girdle infected
shoots. Fruit infection is particularly destructive and can result
in complete crop loss. The fungus survives from one season to the
next in mummified fruit. Infection periods can be identified based
on the occurrence of rainfall and the relationship between temperature and leaf wetness duration.

Botrytis bunch rot, caused by Botrytis cinerea Pers., occurs throughout the viticultural world. Despite its destructive capabilities, B. cinerea, affectionately called the "noble rot" organism, is the prized ingredient in the Auslese, Beerenauslese, and Trockenbeerenauslese wines of Germany and the Sauternes wine of France. The fungus enters fruit through mechanical or other injuries, such as cracking caused by powdery mildew, or directly through the epidermis of ripening berries. Moisture in the form of fog or dew and temperatures of 15-25C are ideal for disease development.

The new fungicides being developed for use on grapes in the US fall into three categories: ergosterol biosynthesis inhibitors acylalanines, and dicarboximides.

Ergosterol Biosynthesis Inhibitors (EBI)

The EBI compounds have many interesting and unique characteristics that allow much more flexibility in control programs than ever before. Some of these unique characteristics are beneficial, but others may be detrimental (Table III).

Table III. Impact of New Fungicides on Grower Management Practices

Characteristics	Beneficial	Detrimental
Effectiveness	X	
Local systemic activity	X	
Vapor activity	X	
Lengthened spray intervals	X	
Low rates per hectare	X	X
After-infection, presymptom uses	X	X
Cost of program	X	X
Narrow spectrum		X
Potential for fungicide resistance		X
Growth regulator effects	X	X
Nontarget effects	X	X
Flexibility	X	
Integrated Pest Management programs	X	

Effectiveness. Several EBI compounds are highly effective against powdery mildew and black rot of grapes.

Local Systemic Activity. The active ingredient(s) enters the plant tissue and moves in the xylem, where it is not subjected to weathering from rainfall. This is a great benefit to growers in the eastern US where summer rainfall is common, and to California growers where occasional unseasonal rains may occur. Problems arise when growers assume these compounds are fully systemic and move in the xylem and phloem, (as does the herbicide glyphosate). Growers must understand that local systemic activity will not rectify poor spray coverage. Furthermore, this activity in relation to protection of berries is unclear.

Vapor Activity. The vapor activity of some EBI compounds is sufficient to control powdery mildew in the greenhouse (13). Grapevine
canopies develop a partially enclosed environment and provide a
logical model system in the transition from greenhouse to field
experiments. During the 1984 growing season, pieces of cheesecloth,
previously soaked in etaconazole (Vangard), CGA 71818 (Topas) or
triadimefon (Bayleton), were wrapped around the trellis wire inside
the grapevine canopy. Etaconazole, CGA 71818 and triadimefon used in
this way resulted in 55%, 87% and 71% control of powdery mildew on
cv. Delaware grape clusters, respectively (Pearson, unpublished
data). Control was less satisfactory on cultivars (such as cv.
Rosette) with upright, open growth habits. The advantages of this
experimental control method are: 1) only one treatment per season
may be needed, 2) improved control of powdery mildew on clusters
inside the canopy may be obtained, and 3) no residues are present on
fruit at harvest.

Lengthened Spray Intervals. Bayleton applied on a 3-4 week schedule
in California for control of powdery mildew and 2-3 week schedule in
the eastern US resulted in savings on fuel, labor, and equipment
compared to the standard sulfur application schedule of 7-10 days
(Table IV).

Table IV. Comparison of Sulfur and Bayleton Powdery Mildew
Control Programs in the Eastern United States and California

	Sulfur		Bayleton 50W	
	East. US	Calif.	East. US	Calif.
Amount product/hectare	4.5 kg (WP)	11.2 kg (dust)	140-210 gm	280-420 gm
Application interval (days)	7-10	7-10	14-21	21-30
Number of applications/season	8-12	6-8	3-6	2-3

Low Rates per Hectare. Bayleton 50WP applied at 140-420 gm per
hectare results in no sprayer nozzle plugging, little visible residue
on table grapes, smaller storage space requirements for the product,
and lighter loads for air applicators; however, the grower must also
be careful in measuring the proper amount of fungicide because excess
product can be expensive in terms of phytotoxicity and cost of material. Furthermore, the use of Bayleton, measured in grams per hectare, results in potentially less pollution of the environment than
when traditional fungicides, measured in kilograms per hectare, are
used.

After-infection, Presymptom Uses. Because of the after-infection
control capabilities of the EBI compounds, the eastern grower can for
the first time control black rot 72-96 hours after infection has
taken place. This approach to controlling black rot can mean substantial savings in the number of applications, especially during dry
or moderately dry years. Unfortunately, there are several drawbacks
to this approach. Generally, growers are not accustomed to monitor-

ing rain events and temperature precisely enough to determine infection periods. Also, it requires purchase and maintenance of relatively expensive weather instruments or an in-field microprocessor-based disease prediction system similar to the Reuter-Stokes apple scab predictor.

The EBI compounds do not appear to be as good protectants as conventional fungicides used for control of black rot. Therefore, a tank mix of Bayleton and a conventional fungicide (e.g. ferbam or mancozeb) is recommended. This combination may be used as follows: the grower waits for an infection period, applies the tank mix within 72-96 hours of its start and then, assumes protection for 7-10 days. Following the 7-10 day period, the grower waits for the next infection period, at which time another spray is applied. On a 2-week spray schedule, the grower would assume 10 days protection and 4 days after-infection control with the subsequent spray. In a wet year, the after-infection program may not save sprays and the grower risks not being able to get into the vineyard in time or being able to cover his entire acreage within the 72-96 hour limit for after-infection control. Furthermore, the spray application must be done carefully to ensure complete coverage of all infection sites, a very difficult task with present technology. Another problem with this approach involves education of the grower regarding the terms after-infection and pre-symptom control. Some growers might wait until they see lesions before they begin spraying. This approach would not only result in failure to control the disease, but would increase the size of the fungal population that is exposed to the chemical, possibly increasing the chances of selection for resistant strains of the fungus.

Cost of Fungicide Program. As noted above, the extended spray intervals for powdery mildew control can mean reduced application costs for the season. In 1983, many California growers were on a season-long Bayleton program, but in light of 1984 economic conditions many growers returned to sulfur dustings or used Bayleton and sulfur alternately. Since most small growers in California own sulfur dusting machines, but few own vineyard sprayers, Bayleton must be applied by custom applicators, thereby adding considerable expense to each application.

Narrow Spectrum. Most conventional fungicides used on grapes protect against at least two diseases, and occasionally against three or four diseases. However, the EBI compounds are generally limited to control of powdery mildew and black rot. Because of this narrow spectrum of activity, the EBI compounds must be tank-mixed with conventional fungicides for control of other diseases, especially downy mildew in the eastern U.S. Furthermore, since conventional fungicides require short spray intervals (7-10 days) to be effective, growers are not able to take full advantage of the extended interval capabilities (14-21 days) of the EBI compounds. They would spray on a 7-10 day schedule, and add Bayleton to every other spray.

Potential for Resistance. Resistance appears to be the major potential obstacle to optimal use of many of the new fungicides. It is clear from the experiences with triazole resistance in powdery mildew

of barley (3) and cucurbits (10) that resistance must also be considered a possibility with the grape powdery mildew fungus. Therefore, strategies for delaying resistance must be implemented from the
outset of their commercial usage. Although the best approach to
avoiding resistance still eludes researchers, growers in New York,
where Bayleton has been used commercially since 1980, are warned not
to use this material on an exclusive, season-long program. We
recommend that they alternate other mildewcides with Bayleton or use
tank mixtures at full rates. Furthermore, it is not recommended that
they use these fungicides alone under high disease pressure as
eradicants. Unfortunately, the threat of resistance may counterba-
lance the desirable attributes of the EBI compounds, notably their
extended interval efficacy and eradicant activity.

Growth Regulator Effects. Some of the EBI compounds have interesting
plant growth regulator (PGR) effects when used at excessive rates,
apparently due to an inhibition of gibberellic acid synthesis (12)
resulting in shortened internodes and small, thick, dark green leaves
(Figure 1). A common problem in grapevine canopy management is
adequate light penetration to the basal nodes of current season's
shoots which will provide next year's crop. There is a direct
relationship between yield from these basal nodes and light exposure
during their formation the previous year. This is not only a problem
in low light areas such as New York, but also in high light areas
such as California where large canopies shade the fruiting wood.
Perhaps the PGR effects of EBI compounds on grape could be beneficial
in that shorter internodes and stiffer shoots with more upright
growth habit would allow better leaf exposure to sunlight and increase node fruitfulness. This growth habit combined with smaller
leaves would also allow better air circulation, better spray penetration and shorter wetting periods resulting in less disease development. Unfortunately, the PGR effects might also result in more
compact clusters with an increased potential for bunch rot problems.

Nontarget Effects. The shift from sulfur dusting to Bayleton spraying in California appears to be associated with an increase in
predatory mite populations, which were uncommon in vineyards on
sulfur programs (personal communication). On the other hand, pesti-
ferous Erineum mites and thrips appear to increase where sulfur usage
has been reduced, although a clear cause and effect relationship has
not been established.

Flexibility. The introduction of Bayleton for control of powdery
mildew in California vineyards has given growers flexibility in their
management programs that sulfur dusting did not allow. When growers
use Bayleton, they need not program their irrigation schedules as
carefully as they did when weekly sulfur applications were required.
Many California grape growers also have stone fruits that are harvested in May. Although May is a critical time for powdery mildew
development in vineyards, extended spray intervals using Bayleton
give the grower more time to devote to the May harvest of tree fruit.
Sulfur is phytotoxic at temperatures above 32C requiring that growers
keep a close watch on the weather at the time sulfur is applied;

Figure 1. Plant growth regulator effects of an EBI compound; note stunted shoots with small, cupped, dark green leaves.

however, the use of Bayleton eliminates this source of potential injury.

Integrated Pest Management (IPM) Programs. The after-infection activity of the EBI compounds against black rot of grape gives growers a chemical tool analogous to those now available to entomologists who, based upon counts of insect or mite populations, can recommend sprays when needed. Previously, without similar chemical tools, plant pathologists have had to rely on protectant fungicides that did not permit full benefit of IPM monitoring programs. By careful monitoring of infection periods it may now be possible to save sprays when not needed or to apply more sprays under severe disease pressure, preventing crop loss.

Acylalanines

Three acylalanine compounds, metalaxyl (Ridomil), cyprofuram (Vinicur) and benalaxyl (Galben), have been tested extensively for control of grape downy mildew in New York. None of these compounds are currently registered in the US. They are narrow spectrum, but highly effective compared to standard protectants such as captan, folpet or mancozeb. They have local systemic activity and possess protectant and antisporulant, as well as after-infection activity. Similar to the EBI compounds, the after-infection capabilities and their activity over extended intervals (14-21 days) make the acylalanines desirable tools for IPM programs.

Unfortunately, resistance to metalaxyl by the grape downy mildew fungus resulted in crop loss in France during 1980 and 1981 (2). Resistance developed in spite of the fact that Ridomil had been sold to grape growers as a package mix with folpet. Since activity over extended intervals is one of the desirable characteristics of Ridomil, it was used as a package mix on a 14-day schedule. The activity of folpet, especially under high disease pressure, lasts only 7-10 days. Therefore, Ridomil remained unaided during the latter half of the 2-week interval. All of the acylalanines are being developed as package mixes (either with folpet or mancozeb) for use on grapes in the US. If spray intervals shorter than 14 days are recommended, it is unlikely growers will be willing to pay for the added cost of the acylalanine when they can get good commercial control of downy mildew with standard protectant fungicides applied alone on a 7-10 day schedule.

An additional problem with most package mixes is the low amount of standard protectant fungicide added to the mix. Not only are the rates borderline to control downy mildew if resistance develops, but they are inadequate for broader disease control. For example, black rot control with Ridomil MZ58 requires that additional mancozeb or another protectant, such as ferbam, be added to the tank mix. Furthermore, a 14-day schedule to control black rot early in the growing season may be inadequate and a grower would need to supplement the Ridomil MZ58 program with additional black rot sprays.

An after-infection approach to controlling downy mildew could be developed, providing growers monitor downy mildew infection periods using the same instrumentation discussed under black rot. Acylalanine tank-mixed with folpet or mancozeb would be applied after a downy mildew infection period was identified. No further sprays would be applied for 14 days (assume adequate protection) after which a spray would be applied following the next infection period. In dry seasons, this approach would undoubtedly reduce the number of sprays, but perhaps not in wet seasons. Unfortunately, this program is vulnerable to the selection of acylalanine resistant strains if growers become careless and wait until they see sporulating lesions.

Dicarboximides

The third group of new fungicides to be tested on grapes in the US is the dicarboximide group. Vinclozolin (Ronilan) and iprodione (Rovral) are the members of this group that have been extensively tested for control of Botrytis bunch rot.

Our experience with benzimidazole resistance in the powdery mildew fungus (8) and Botrytis (7), in addition to reports from Europe (5, 11) and Canada (6) concerning resistance to the dicarboximides by Botrytis, has influenced our recommendations regarding the use of dicarboximides on grapes in New York. Our current recommendation is to apply two sprays: 1) when the first berries reach 5% sugar and 2) two weeks after the first application. Several years of testing in New York (9) indicated that additional applications of the dicarboximides at bloom and bunch-closing did not improve control of Botrytis bunch rot at harvest. The rationale for this program is to keep the cost of these expensive applications as low as possible and to reduce the selection pressure against the population by applying as few sprays as possible (hopefully delaying the development of resistant strains), yet using them at the most optimal time.

Summary

The short-term outlook for grape growers is rather dim. Current economic pressures are forcing grape growers across the nation to cut corners. Perhaps the most shortsighted cost cutting step is the reduction in use of fungicides. In New York during 1984, for example, many growers reduced the number of sprays by 50%, and some applied only one or two sprays. In California, many growers who were on full Bayleton schedules in 1983 went back to sulfur or used only one or two Bayleton sprays in 1984.

Reduced spray inputs mean an even greater temptation by growers to wait until they see disease and then to use the new fungicides as eradicants under heavy disease pressure. We must be concerned about this type of usage because of the threat of resistance. Obviously for the short term, growers will continue to reduce inputs in an effort to survive economically. The challenge to extension and research plant pathologists is to develop fungicide use strategies that avoid resistance yet are responsive to growers' economic constraints. Obviously, the challenge to the chemical industry is to develop new fungicides with all the great attributes of the EBI compounds, but without the threat of resistance.

Literature Cited

1. Annon. 1982. Situation de la viticulture dans le monde en 1981. Bull. de L' O.I.V. 55:801-834.
2. Clerjeau, M. and J. Simone. 1982. Apparition en France de souches de mildiou (Plasmopara viticola) résistantes aux fongicides de la famille des anilides (Métalaxyl, Milfurame). Le Progres Agricole et Viticole 99:59-61.
3. Fletcher, J. T. and M. S. Wolfe. 1981. Insensitivity of Erysiphe graminis f. sp. hordei to triadimefon, triadimenol and other fungicides. Proc. Brit. Crop Prot. Conf. (Brighton, 1981) 2:633-640.
4. Himelrick, D. G. 1984. The potential for table grape production in New York State. Proc. N.Y.S. Hortic. Soc. 129:51-62.
5. Holz, B. 1979. Über eine Resistenzerscheinung von Botrytis cinerea an Reben gegen die neuen Kontaktbotrytizide im Gebiet der Mittelmosel. Weinberg und Keller 26:18-25.
6. Northover, J. and J. A. Matteoni. 1984. Iprodione-resistant Botrytis cinerea found in Ontario vineyards and greenhouses. Phytopathology 74:810 (Abstr.).
7. Pearson, R. C., D. A. Rosenberger, and C. A. Smith. 1980. Benomyl-resistant strains of Botrytis cinerea on apples, beans and grapes. Plant Disease 64:316-318.
8. Pearson, R. C., and E. F. Taschenberg. 1980. Benomyl-resistant strains of Uncinula necator on grapes. Plant Disease 64:677-680.
9. Pearson, R. C., and D. G. Riegel. 1983. Control of Botrytis bunch rot of ripening grapes: Timing applications of the dicarboximide fungicides. Am. J. Enol. Vitic. 34:167-172.
10. Schepers, H. T. A. M. 1983. Decreased sensitivity of Sphaerotheca fuliginea to fungicides which inhibit ergosterol biosynthesis. Neth. J. Pl. Path. 89:185-187.
11. Schüepp, H., M. Küng, and W. Siegfried. 1982. Developpement des souches de Botrytis cinerea résistantes aux dicarboximides dans les vignes de la Suisse alémanique. Bull. OEPP 12:157-161.
12. Siegel, M. R. 1981. Sterol-inhibiting fungicides: Effects on sterol biosynthesis and sites of action. Plant Disease 65:986-989.
13. Szkolnik, M. 1983. Unique vapor activity by CGA-64251 (Vangard) in the control of powdery mildews roomwide in the greenhouse. Plant Disease 67:360-366.
14. Wines & Vines, February 1984, p. 39-40.
15. Wines & Vines, July 1984, p. 37-39.
16. Wines & Vines, July 1984, p. 50-51.

RECEIVED October 1, 1985

11

Advances in Fungicide Chemistry and Fungal Control
Summary and Comments

Hugh D. Sisler

Department of Botany, University of Maryland, College Park, MD 20742

This publication comprises papers dealing with factors which constitute the basis for the advancement of chemical control of fungal diseases. This summary provides comment on some of the points made in these papers and also presents additional information from other sources as well as some personal opinions and ideas.

Professor Büchel dealt with the history of azole chemistry and the impact of azole fungicides on the control of fungal pathogens of plants and humans. This presentation made a number of valuable points of interest to organic chemists and biologists. One significant point made by Professor Büchel concerns the wide structural variations permissible in the lipophilic substituent at position 1 of azole fungicides without loss of the basic antifungal activity. This substituent, nevertheless, dictates biochemical and biological specificity as well as properties which determine practical success of these fungicides. In a potent fungicide, this substituent apparently must have a configuration which permits the interaction of an N atom of the azole group with a haem iron atom (or with some other group) of an enzyme while binding tightly to the enzyme to reinforce the linkage of the N atom to the iron. As pointed out by Professor Büchel, the possibilities for designing azole compounds of differing biological activity are almost limitless. One serious challenge which may be encountered in fungicide synthesis is that of designing structures which do not seriously interfere with any of the various mammalian cytochrome P-450 enzymes.

Dr. Berg focused attention on the biochemical specificity of fungicides acting in the ergosterol biosynthetic pathway. He pointed out that, while many ergosterol biosynthesis inhibiting (EBI) fungicides, including the azoles, primarily block sterol C-14 demethylation, the fungicide tridemorph blocks $\triangle^8 \longrightarrow \triangle^7$ isomerization or reduction of the sterol C-14, 15 double bond. Another fungicide, a 15-azasterol antibiotic, is a potent inhibitor of $\triangle^{14}$ sterol reductase with a Ki of 2nM (1). New fungicides acting in the sterol biosynthetic pathway at yet another site have been reported recently. Allylamine derivatives of the naftifine group are fungicides which block squalene expoxidation. The derivative SF 86-327, for example,

0097-6156/86/0304-0157$06.00/0
© 1986 American Chemical Society

is a potent inhibitor of squalene epoxidase with an inhibition constant of 3×10^{-8}M (2). The large number of fungicides acting on ergosterol biosynthesis indicate that this pathway is a desirable one to target for fungitoxic action. However, the question might be raised as to whether inhibition of ergosterol biosynthesis alone is primarily responsible for antifungal action of all of the compounds reported to act in the pathway. A good argument might be made that this is the case for the 15-azasterol antibiotic since its toxicity can be reversed with exogenous ergosterol (3). However, toxicity of sterol C-14 demethylation inhibitors is not reversed by adding exogenous ergosterol, leaving some doubt that inhibition of ergosterol biosynthesis alone is responsible for toxicity. Perhaps we should look beyond the sterols and into steriod hormone biosynthesis and function for a better understanding of the fungitoxic action of EBI compounds. The pathway from lanosterol to ergosterol may be of less specific importance for providing appropriate sterol stuctures for membranes than it is as a segment of a pathway for steroid hormone biosynthesis. The recent discovery that Saccharomyces cerevisiae produces an estrogen binding protein (4) and produces an estrogenic substance (5) should intensify interest in the possibility that fungi produce steroid hormones and that EBI fungicides might block their synthesis or action. It has, for example, been shown that testosterone and progesterone partially reverse toxicity of EBI inhibitors in some fungi (6). Of interest is the demonstration that ketoconazole, a sterol C-14 demethylation inhibitor, not only blocks testosterone production in mammalian systems but also displaces steroid hormones from serum transport proteins (7). Cytochrome P-450 enzymes are very prominent in the biosynthesis of steroid hormones from desmethyl sterols and action of an EBI on these enzymes could explain why fungitoxicity in some cases is not reversed by the addition of ergosterol.

Professor Führ discussed penetration, translocation and distribution of fungicides in plants. These aspects of fungicide performance are often critical in determining the success of a systemic fungicide in a particular type of application. The ability to penetrate into plant tissue and move while retaining activity therein is the primary basis for the superiority of the newer systemic fungicides over the older surface protectants. Internal therapeutants may also influence the host physiology and as a consequence, be assisted by the natural defense system of the host.

Dr. Scheinpflug addressed this point in his presentation. He presented light and electron microscopic evidence which suggests that some systemic fungicides may induce host resistance to an invading fungal pathogen. The roles of direct fungitoxicity and activated host resistance systems in such cases is difficult to resolve. By simply slowing down the rate of growth of a pathogen, a compound might provide the host plant the time necessary to mobilize defense systems and thus greatly enhance the apparent effectiveness of the compound as a fungicide.

Professor Schwinn discussed new advances in the chemical control of plant parasitic Oomycetes and Peronosporales. Some of the most devastating foliar and root diseases are caused by these fungi, but they are often not controlled by systemic fungicides which control

pathogens of other mycological groups. Cymoxnil, acylalanines and
phosethyl-Al are some of the newer fungicides specifically effective
for control of Oomycetes and Peronosporales. The acylalanines have
proven to be highly active fungicides but have experienced some
resistance problems which hopefully can be managed by proper anti-
resistant strategies.

Professor Dekker discussed non-fungicidal compounds which
control disease by increasing host resistance or by decreasing the
ability of the pathogen to attack the host. The dichlorocyclopropane
carboxylic acids and probenazole (a saccharine relative) are examples
of compounds which sensitize the host plant to respond in a resistant
manner. Melanin biosynthesis inhibitors such as tricyclazole,
pyroquilon and fthalide are representatives of compounds which
decrease pathogenic capabilities of the fungus. Although some of the
aforementioned compounds such a probenazole and tricyclazole have
been adopted for practical control of plant diseases, representatives
in this category are few in number and of relatively minor importance. The idea of controlling plant diseases by manipulation of
host-resistance systems or pathogenic mechanisms in the fungus is
attractive. The high specificity of many host/parasite relationships
suggest that recognition may often be the most critical factor in the
initiation of a rapid resistant response in the host. These interactions may be difficult to regulate chemically without injury to the
host plant. A much better understanding of the factors involved is
necessary before a rational chemical manipulation of these interactions can be utilized as a means of disease control. Conventional
screening holds only very limited promise for discovering effective
chemicals of this category; however, any compounds discovered may be
valuable in helping to elucidate host/parasite relationships and thus
promote further advances.

Professors Hoffman, Cole, Jones and Pearson discussed the use of
new fungicides in various crops with respect to increased crop
yields, economics, integrated pest management (IPM) systems, epidem-
ology and fungal resistance to fungicides. The eradicative action of
some of the newer fungicides permits post-infection application and
this feature makes the compounds particularly desirable for IPM
systems. In some cases, the fungicides are less effective in pre-
infection applications than in post-infection applications. This is
apparently due to the loss of fungicide to weathering and ultraviolet
degradation which occurs when the fungicide is present prior to the
infection period. Fungal resistance to desirable new types of
fungicides continues to be a serious concern. The best apparent way
to combat this problem is to develop use strategies which will
decrease the possibility for selection of resistant strains. This
will require close cooperation between industry, state or government
institutions, and the growers. There is need for new fungicides with
modes of action differing from those of compounds now in use in order
to design application sequences that minimize overuse of chemicals in
one mode of action category.

The future holds considerable promise for the development of new
systemic fungicides and compounds which modify host-resistance or
fungal pathogenicity. Among the major obstacles which will be

encountered are the problems of fungal resistance and increasingly
rigid toxicological standards.

Literature Cited

1. Bottema, C.K.; Parks, L.W. Biochim. Biophys. Acta 1978, 531,
    301-7.
2. Petranyi, G.; Ryder, N.S.; Stütz, A. Science 1984, 224, 1239-
    41.
3. Taylor, F.R.; Rodriguez, R.J.; Parks, L.W. Antimicrob. Agents
    Chemother. 1983, 23, 515-21.
4. Feldman, D.; Do, Y.; Burshell, A.; Stathis, P.; Loose, D.S.
    Science 1982, 218, 297-8.
5. Feldman, D.; Stathis, P.A.; Hirst, M.A.; Stover, E.P.; Do, Y.S.
    Science 1984, 224, 1109-11.
6. Sherald, J.L.; Ragsdale, N.N.; Sisler, H.D. Pesti. Sci. 1973,
    4, 719-27.
7. Grosso, D.S.; Boyden, T.W.; Pamenter, R.W.; Johnson, D.G.;
    Stevens, D.A.; Galgiani, J.N. Antimicrob. Agents Chemother.
    1983, 23, 207-12.

RECEIVED January 8, 1986

The Division of Pesticide Chemistry

of the

American Chemical Society

Commemorates by this certificate
the fifteenth presentation of the

Burdick and Jackson International Award
for Research in Pesticide Chemistry,
sponsored by Burdick and Jackson Laboratories, Inc.,
Muskegon, Michigan

to

KARL HEINZ BÜCHEL

for outstanding contributions to the field.   His major
research achievements have involved the rational development
of new pesticide chemicals based on biological
activity/physicochemical parameter correlations and on
planned interference with specific biochemical processes.

Presented at the 188th National Meeting
of the
American Chemical Society

Philadelphia, Pennsylvania

August 27, 1984

Dr. Robert M. Hollingworth (left), the 1984 Chairman of the
Division of Pesticide Chemistry of the American Chemical Society,
presents to Prof. Karl Heinz Büchel (right) an award plaque
commemorating his selection as the 1983 recipient of the Burdick
and Jackson International Award for Research in Pesticide
Chemistry.

Presentation of the 1983 Burdick and Jackson International Award for Research in Pesticide Chemistry to Prof. Karl Heinz Büchel by Robert M. Hollingworth, Chairman of the Division of Pesticide Chemistry of the American Chemical Society.

As chairman of the Division of Pesticide Chemistry it is my special pleasure and privilege to introduce this year's recipient of the Burdick and Jackson International Award in Pesticide Chemistry. This award is given annually for "outstanding contributions to research in pesticide chemistry" and represents one of the highest honors that the agrochemicals community can bestow on one of its members. We are here today to present this honor to our distinguished colleague, Prof. Karl Heinz Büchel, Director of Central Research with Bayer AG.

Prof. Büchel studied chemistry at the Chemical Institute of the University of Bonn from 1952 to 1958 at which date he received his Doctorate of Science, and he continued there as an assistant to Prof. F. Korte from 1958 to 1961. Incidentally, it is a pleasure to welcome Prof. Korte to this gathering today to celebrate the remarkable success of his former student. From Bonn, Prof. Büchel moved to the Research Institute of Shell International at Birlinghoven for 5 years as a department head before beginning his career with Bayer AG. In the short space of the 10 years that followed, he rose from being a member of the Central Research Laboratory at Leverkusen to the position of Director of Central Research for Bayer and the Speaker for Research on Bayer's Board of Management.

The 120 scientific publications and 238 patent applications of which Prof. Büchel is either author or co-author testify to his personal ability and achievements in research in pesticide chemistry and to his skills in supervising and directing the research efforts of others. His research has covered many topics, too numerous to present in full here, but covering all facets of pesticidal activity. In particular, his name is associated with the discovery of the aminotriazinone herbicides--most notably metribuzin and metamitron. He has supervised the development of the insecticide, plifenate and the acaricide, azocyclotin. His work on the mode of action and structure-activity relationships of photosynthesis inhibitors and uncouplers of mitochondrial oxidative phosphorylation has been outstanding and, I might add, of considerable interest to me personally. However, his greatest achievement to date has been in pioneering the discovery and development of the azole fungicides that are the topic of today's symposium. As we shall see, this work begun in 1966 has

led to a very valuable and extensive series of commercial
products active against human, animal, and plant
pathogenic fungi and, in the case of the plant
fungicides, often showing an excellent degree of systemic
action. It is notable that these discoveries were made
and optimized by a rigorous scientific approach involving
planned interference with specific biochemical processes
and quanitative correlations of biological activity with
physicochemical characteristics in the active series.

While generating these remarkable discoveries in
industry, Prof. Büchel has maintained close connections
with the academic world, including lecturing in organic
chemistry at the Technical University of Aachen where he
was appointed as a Professor in 1975. In 1981, he was
awarded an Honorary Doctorate of the Faculty of
Agriculture and Horticulture of the Technical University
of Munich. Nor has he neglected other services to
chemistry beyond his work with the Bayer Company. He is
the Chairman of the Board of Trustees of the German
Chemical Information Service, Chairman of the Committee
on Chemistry and Industry of the International Union of
Pure and Applied Chemistry, and a member of the Executive
Committee of the Society of German Chemists, among
numerous other activities. To cap off this list of
achievements, he is the author of one book on pesticides,
"Crop Protection and Pest Control" (1977), and the editor
of a second, "Pesticide Chemistry" (1983), which have
been widely recognized and appreciated.

Today we are honoring Prof. Büchel, but on behalf of
our Division, I also take pleasure in noting that this
award goes to a scientist from the industrial sector of
the agrochemicals field, and particularly to someone who
represents the German branch of this industry, which has
made so many outstanding contributions to pesticide
discovery in the last century. Recognition of this
contribution was overdue on our part, and I am glad to
see the omission rectified, at least in part, today. It
is also gratifying to see that so many of our friends
from Europe were able to join us for this presentation
and symposium. We take great pleasure in their company
and appreciate the honor they do us and Prof. Buchel by
their presence.

The International Award has been given on fourteen
previous occasions, but I am sure that having heard the
record of Prof. Büchel's achievements, you will agree
with me that we have not had a more deserving recipient
than on this occasion.

Acceptance of the 1983 Burdick and Jackson International Award by Professor Karl Heinz Büchel.

Mr. President, Ladies, and Gentlemen:

It is with great pleasure and much gratitude that I accept the Burdick and Jackson Award for Contributions to Pesticide Chemistry, particularly in view of the fact that for many years my scientific work has been dedicated to the chemistry of crop protection. I am proud that this great honor has been awarded by such a prestigious organization as the Pesticide Division of the American Chemical Society.

This award for scientific achievement also obliges the recipient to record his appreciation and give thanks to the people who introduced him to the scientific and professional world. First of all, there is my university teacher and doctorate supervisor at Bonn University, Prof. Dr. Friedhelm Korte, whose versatile and multi-faceted school, with its many interdisciplinary activities, taught me very much.

My early professional years were spent in Shell International Research, and in those days we had very close collaboration with the Shell Laboratories in the United States. During that time, I had the pleasure of receiving important educational support from scientists in this country. I would like particularly to mention Dr. Barney Soloway and Dr. Sheldon Lambert from Shell Modesto Laboratories and acknowledge them for their pioneering and innovative ideas on structure-activity problems.

In my first years with the Bayer Research and Development organization, I particularly enjoyed support from Prof. Wegler, then Director of Research, and Dr. Risse, then Head of the Agrochemicals Division. Both gentlemen gave me not only the freedom, but also a budget to follow my ideas in a sometimes individual way.

A chemist who devotes himself to the development of new biologically active chemicals is swiftly taught modesty. Success in this area is always the product of interdisciplinary cooperation. Please allow me to emphasize that I also accept this great honor in the name of all my colleagues, not only for the ones in Chemistry, but also in Biology, the Agrosciences, Medicine, and Toxicology. This honor and my thanks are also due to them.

I accept this ACS Award as a recognition of the 90-year tradition of agrochemical research of the Bayer Company and the many pioneering achievements of this organization for the benefit of agriculture.

In my following presentation, I would like to illustrate a special chapter of these contributions--"The History of Azole Fungicides". In the background of this story you may find some indication of the ingredients for successful research: a bright idea, freedom of activity, flexibility, patience, persistence, and last, but not least, good fortune.

Thank you!

Professor Director Doctor
Karl Heinz Büchel
Bayer AG
Forschung und Entwicklung
509 Leverkusen Bayerwerk
Federal Republic of Germany

Author Index

Subject Index

Production and indexing by Susan F. Robinson
Jacket design by Pamela Lewis

Elements typeset by Hot Type Ltd., Washington, DC
Printed by The Sheridan Press, Hanover, PA
Bound by Maple Press Co., York, PA

RECENT ACS BOOKS

"Chromatography and Separation Chemistry:
Advances and Developments"
Edited by Satinder Ahuja
ACS SYMPOSIUM SERIES 297; 304 pp; ISBN 0-8412-0953-7

"Natural Resistance of Plants to Pests"
Edited by Maurice B. Green and Paul A. Hedin
ACS SYMPOSIUM SERIES 296; 244 pp; ISBN 0-8412-0950-2

"Nutrition and Aerobic Exercise"
Edited by Donald K. Layman
ACS SYMPOSIUM SERIES 294; 150 pp; ISBN 0-8412-0949-9

"The Three Mile Island Accident: Diagnosis and Prognosis"
Edited by L. Toth, A. Malinauskas, G. Eidam, and H. Burton
ACS SYMPOSIUM SERIES 293; 302 pp; ISBN 0-8412-0948-0

"Environmental Applications of Chemometrics"
Edited by Joseph J. Breen and Philip E. Robinson
ACS SYMPOSIUM SERIES 292; 286 pp; ISBN 0-8412-0945-6

"Desorption Mass Spectrometry: Are SIMS and FAB the Same?"
Edited by Philip A. Lyon
ACS SYMPOSIUM SERIES 291; 248 pp; ISBN 0-8412-0942-1

"Catalyst Characterization Science:
Surface and Solid State Chemistry"
Edited by Marvin L. Deviney and John L. Gland
ACS SYMPOSIUM SERIES 288; 616 pp; ISBN 0-8412-0937-5

"Polymer Wear and Its Control"
Edited by Lieng-Huang Lee
ACS SYMPOSIUM SERIES 287; 421 PP; ISBN 0-8412-0932-4

"Ring-Opening Polymerization:
Kinetics, Mechanisms, and Synthesis"
Edited by James E. McGrath
ACS SYMPOSIUM SERIES 286; 398 pp; ISBN 0-8412-0926-X

"Multicomponent Polymer Materials"
Edited by D. R. Paul and L. H. Sperling
ADVANCES IN CHEMISTRY SERIES 211; 354 pp; ISBN 0-8412-0899-9

"Formaldehyde: Analytical Chemistry and Toxicology"
Edited by Victor Turoski
ADVANCES IN CHEMISTRY SERIES 210; 393 pp; ISBN 0-8412-0903-0

For further information contact:
American Chemical Society, Sales Office
1155 16th Street NW, Washington, DC 20036
Telephone 800-424-6747